T0214515

Lecture Notes
in Business Information Processing

357

More information about this series at http://www.springer.com/series/7911

Jennifer J. Xu · Bin Zhu ·
Xiao Liu · Michael J. Shaw ·
Han Zhang · Ming Fan (Eds.)

The Ecosystem of e-Business: Technologies, Stakeholders, and Connections

17th Workshop on e-Business, WeB 2018
Santa Clara, CA, USA, December 12, 2018
Revised Selected Papers

Springer

Editors
Jennifer J. Xu ⓘ
Bentley University
Waltham, MA, USA

Xiao Liu
University of Utah
Salt Lake City, UT, USA

Han Zhang ⓘ
Georgia Institute of Technology
Atlanta, GA, USA

Bin Zhu
Oregon State University
Corvallis, OR, USA

Michael J. Shaw
University of Illinois
Urbana-Champaign, USA

Ming Fan ⓘ
University of Washington
Seattle, WA, USA

ISSN 1865-1348   ISSN 1865-1356 (electronic)
Lecture Notes in Business Information Processing
ISBN 978-3-030-22783-8   ISBN 978-3-030-22784-5 (eBook)
https://doi.org/10.1007/978-3-030-22784-5

This Springer imprint is published by the registered company Springer Nature Switzerland AG
The registered company address is: Gewerbestrasse 11, 6330 Cham, Switzerland

Preface

The Workshop on e-Business (WeB) is a premier annual conference on e-business and e-commerce. The purpose of WeB is to provide an open forum for researchers and practitioners to discuss findings, novel ideas, success stories and lessons learned, to map out major challenges, and to collectively chart future directions for e-business. Since 2000, WeB has attracted valuable and novel research that covers both the technical and managerial aspects of e-business. The 17th Annual Workshop on e-Business (WeB 2018) was held in Santa Clara, California, USA, on December 12, 2018.

The theme of WeB 2018 was "The Ecosystem of e-Business: Technologies, Stakeholders, and Connections." The pace of technology-enabled business innovations has been accelerating. The emerging technologies and new business models have not only transformed traditional e-business firms and markets, but have also changed both the roles of stakeholders and the connections in business networks. New technologies such as data analytics and artificial intelligence have been applied to enhance the efficiency and effectiveness of e-business transactions and to facilitate better decision-making. Disruptive technologies such as blockchain have the potential to overturn the way in which business is being conducted, whereas the sharing economy is impacting global markets. All of these changes, in turn, will also shape the ecosystem of e-business.

WeB 2018 provided an opportunity for scholars and practitioners to exchange ideas and share findings on these topics. Original research articles addressing a broad coverage of technical, managerial, economic, and strategic issues related to consumers, businesses, industries, and governments were presented at the workshop. These articles employed various IS research methods such as survey, analytical modeling, experiments, computational models, and design science.

Among 41 papers presented at WeB 2018, 19 papers were selected to be published in this volume of LNBIP. We would like to thank all the reviewers for their time and effort and for completing their review assignments on time despite tight deadlines. Many thanks to the authors for their contributions.

May 2019

Jennifer Xu
Bin Zhu
Xiao Liu
Michael J. Shaw
Han Zhang
Ming Fan

Organization

Honorary Chairs

Hsinchun Chen	University of Arizona, USA
Andrew B. Whinston	University of Texas at Austin, USA

Conference Chair

Michael J. Shaw	University of Illinois at Urbana-Champaign, USA

Organizing Co-chairs

Bin Zhu	Oregon State University, USA
Jennifer Xu	Bentley University, USA
Xiao Liu	University of Utah, USA
Ming Fan	University of Washington, USA
Han Zhang	Georgia Institute of Technology, USA

Local Organizing Committee Chair

David Zimbra	Santa Clara University, USA

Program Committee

Reza Alibakhshi	HEC Paris, France
Joseph Barjis	San Jose State University, USA
Hsin-Lu Chang	National Chengchi University, Taiwan
Michael Chau	The University of Hong Kong, SAR China
Cheng Chen	University of Illinois at Chicago, USA
Rui Chen	Iowa State University of Science and Technology, USA
Ching-Chin Chern	National Taiwan University, Taiwan
Muller Cheung	Hong Kong University of Science and Technology, SAR China
Huihui Chi	ESCP Europe, France
Aidan Duane	Waterford Institute of Technology, Ireland
Samuel Fosso	Toulouse Business School, France
Henry Han	Fordham University, USA
Wencui Han	University of Illinois at Urbana-Champaign, USA
Lin Hao	University of Notre Dame, USA
Jinghua Huang	Tsinghua University, China
Seongmin Jeon	Gachon University, South Korea
Chunghan Kang	Georgia Institute of Technology, USA

Contents

Social, Policy, and Privacy Issues

Impact of Social Media on Real Estate Sales

Hui Shi[1]([⊠]), Zhongming Ma[1], Dazhi Chong[2], and Wu He[3]

[1] California State Polytechnic University, Pomona, USA
huishi@cpp.edu
[2] California Lutheran University, Thousand Oaks, USA
[3] Old Dominion University, Nofrolk, USA

Abstract. More and more businesses are using social media to promote services and increase sales. This paper explores the impact of Facebook on real estate sales. First, we examine how Facebook activities are associated with real estate sales. Then, we include time lags in our analysis, because a time lag can be expected between the activates on Facebook and a resulting real estate transaction. The results suggest that: (1) The total numbers of Facebook Likes, links, and stories are positively associated with real estate sales; (2) The sentiment score of Facebook posts is negatively associated with real estate sales; (3) Time lag affects the impact of Facebook activities on real estate sales. The results reveal the predicting value of social media and the power of selected Facebook variables on real estate sales. The research findings can be used to promote sale and forecasting.

Keywords: Decision making · Social media · e-Business · Sentiment analysis · Prediction

1 Introduction

Social media has profoundly changed our social lives and how we communicate with others [1]. Consumers from distinct backgrounds are rapidly adopting social media sites to communicate with friends and members of society and enhance their social lives [2]. Consumers who participate in social networks are feeling more empowered in their interactions with e-Businesses. After seeing the potential business value of social media platforms, many companies and e-Businesses have adopted social media sites to increase sales and revenues, increase customer loyalty and retention, create brand awareness and build reputation. For example, Ford Motor Company promoted the release of their new model Ford Focus via Facebook, Twitter and other social media sites [3].

Facebook, Twitter, YouTube, LinkedIn, Pinterest, Instagram and WhatsApp are often considered as most popular social media applications [4]. These social media applications rely mostly on user-generated contents including texts, photos, and videos. Companies can research the large amount, frequently updated social media data to mine useful knowledge, such as customer traffic, correlation between customer comments and sales. Such knowledge can be used in the decision making process later to improve customer satisfaction and increase sales [4, 5].

J. J. Xu et al. (Eds.): WEB 2018, LNBIP 357, pp. 3–14, 2019.
https://doi.org/10.1007/978-3-030-22784-5_1

Recent years, real estate companies and agents have adopted social media sites, such as Facebook business page for advertising. A Facebook business page is an excellent way to attract new clients, promote business, and get feedback from clients. However, to our best knowledge, we have not found any research on finding the impact of social media sites on real estate sales. Therefore, on one side, real estate companies and agents are uncertain about the impact of Facebook business page and Facebook activities on real estate sales. On the other side, homebuyers are uncertain about if they can use Facebook activities to predict the real estate sale trend. To this end, we are interested to address the following research questions:

(1) Is there a correlation between Facebook activities and real estate sales?
(2) To what extent are Facebook activities associated with real estate sales?
(3) To what extent can time lags affect the impact of Facebook activities on real estate sales?

The contributions of this paper are as follows: we investigate how different Facebook activities are associated with real estate sales; we examine how time lags affect the Facebook impact on the sales. This paper is organized as follows. In Sect. 2, we present an overview of relevant literature and theories, and discuss hypothesis development. We then describe research methodology including research model, data collection, and methods in Sect. 3. Subsequently, we analyze and discuss the results in Sect. 4. Finally, we discuss findings, implications and future research in Sect. 5.

2   Theories and Hypothesis Development

2.1   Using Social Media to Predict Sales

Forecasting sales is important in marketing and business. Social media is a form of collective wisdom. Asur and Huberman [6] demonstrate how to utilize sentiments extracted from Twitter to forecast box-office revenues for movies. Gruhl et al. [7] discuss how to use sales rank values and correlating postings in blogs, media and web pages to predict spikes in sales rank. Bollen, Mao and Zeng [8] first analyze the content of daily Twitter feeds by two mood tracking tool, track the changes in public mood state using large-scale Twitter feeds, and explore if public mood correlates to stock market. Bartov, Faurel and Mohanram [9] found that the aggregate opinion in individual tweets successfully predicts a firm's quarterly earnings and announcement returns after exploring a sample set from 2009 to 2012. Joshi and others [10] use the text of film reviews from different sources, movie metadata and linear regression to predict the opening weekend gross earnings. Wu and Brynjolfsson [11] report that a house search index, which is based on search activities from Google search engine, is strongly correlated with future home sales and prices. Schoen and others [12] classify forecasting models into three types: prediction marking models, survey models, and statistical models. And in practice, it is very common to apply statistical models to analyze social media data and make prediction.

2.2   Sentiment Analysis in Social Media

Sentiment Analysis is also known as Opinion Mining, referring to contextual mining of text which identifies and extracts subjective information in the text [13]. As the volume of social media data has been growing massively, mining user-generated content from social media has been a growing interest to obtain users' opinion. Although most of sentiments are simply classified into limited categories such as positive and negative, sentiment analysis is often used to understand the attitude of customers on specific topics or events [13, 14]

There is a large body of work on Sentiment Analysis. Generally, three techniques are used in Sentiment Analysis: machine learning method, lexicon-based method, and hybrid method [15]. The machine learning method applies the existing machine learning algorithms using linguistic features. The lexicon-based method employs a collection of known opinion words. The hybrid method combines the above two methods. Machine learning method includes supervised learning [16] and unsupervised learning [17]. With supervised learning methods, social posts (either document or sentences) can be classified into three categories, positive, negative, and neutral. Any existing supervised learning method can be applied to sentiment classification, such as, naive Bayesian classification [18], and support vector machines (SVM). Unsupervised learning is usually applied when there is no labeled training data.

Twitter and Facebook are common social media platforms used by many sentiment analysis applications [19]. Nakov et al. [20] discuss Sentiment Analysis in Twitter Task. This Task consists of five parts: (1) predicting if a tweet is positive, negative or neutral; (2) predicting whether a tweet conveys a sentiment towards a given topic; (3) estimating the tweet sentiment on a five-point scale from Highly Negative to Highly Positive; (4) estimating the distribution of a set of tweets in Positive and Negative classes; (5) estimating the distribution of a set of tweets in five classes. Severyn and Moschitti [21] apply deep learning in sentiment analysis of tweets. They propose a convolutional neural network for sentiment classification. Saif et al. [22] present SentiCircles, which is a lexicon based method for sentiment analysis on Twitter. SentiCircles can detect sentiment at both entity-level and tweet-level. A novel meta-heuristic clustering method based on K-means and cuckoo search is proposed by Pandey, Rajpoot, and Saraswat [23]. The proposed method finds the optimum cluster-heads from the sentimental contents of tweets. Ortigosa and his colleagues developed an application called SentBuk to extract sentiment from Facebook messages and support sentiment analysis in Facebook [24]. Meire and others [25] discuss the additional value of information available before (leading) and after (lagging) the focal post's creation time in Facebook sentiment analysis. Sentiment analysis has also been used to analyze other social media platforms such as YouTube [26, 27] and has often been used in application areas such as finance and product review [28, 29].

2.3   Hypothesis Development

The number of Facebook Likes increases when you click the "Like" below a post. It is a way to let people know that you enjoy it without leaving a comment. The Facebook Likes has been used in regression analysis to explore its impact on sales and also to forecast sales [30, 31]. According to prior literature, Facebook Likes can drive traffic, induce social selling, and increase sales. Many customers rely on Facebook Likes to decide if they buy a particular product [4, 31, 32]. Building upon the prior literature, we hypothesize that the Facebook Likes on a real estate firm's business page is positively associated with its real estate sales. In another word, if a real estate firm has more Facebook Likes, it tends to get more real estate transactions. This leads to the first hypothesis:

H1. The number of Facebook Likes has a positive influence on real estate sales.

As the volume of social media data has been growing massively, mining user-generated content from social media has been a growing interest to obtain users' opinions. Customer sentiment analysis is a process of gathering customer opinions [33, 34]. Sentiment analysis helps calculate emotions related to a business, product, or brand. For example, sentiment analysis can be used to analyze blog posts to provides useful insights towards a particular topic or product [35]. Sentiment analysis helps marketers to make future plans by analyzing customer reviews and monitors customer's dissatisfaction. On the basis of the prior literature, if a real estate firm's Facebook page has more positive posts, it tends to attract more home clients to work with it. In another word, if a real estate firm has more positive Facebook posts, it tends to have more real estate transactions. This leads to the second hypothesis:

H2. The sentiment of Facebook posts has a positive influence on real estate sales.

Seven other Facebook variables including the total number of posts, total number of photos, total number of videos, total number of comments made under posts, total number of sharers, total number of links, and total number of stories, have been employed in past research [4, 30–32]. The main purpose of a multiple regression is to explore more about the relationship between several independent variables (e.g., Facebook variables) and a dependent variable (e.g., real estate sales) [36]. Multiple regression can find out how different Facebook variables impact real estate sales [30]. Since more customers rely on social media to determine what product they want to purchase, social marketing is now the driving force behind brand awareness, customer engagement and sales. On the basis of prior literature, we assume a relationship between Facebook activities and real estate sales. This leads to the third hypothesis regarding a combinational impact of various Facebook activities:

H3. The activities on Facebook has a positive influence on real estate sales.

Social media sentiments affect sales with a delay, which is named as time lag [37]. Kotler states that consumers go through five stages when buying a product: (1) consumer's recognition of a need or problem; (2) information search; (3) evaluation of alternatives; (4) actual purchase decision; and (5) post purchase behavior [38].The time lag between an increase in positive/negative comments and an increase/decrease in sales is a variable since social media may influence the consumer at any stage (early or late) of the buying process [3]. Similarly, the time lag between activities on Facebook and an increase/decrease in real estate sales should be considered as well. This leads to the fourth hypothesis:

H4. The time lag affects the impact of Facebook activities on real estate sales.

3 Methodology

In order to test the research hypotheses and answer the research questions, we need to explore how the opinions on social media impact real estate sales. In this section, we discuss how we collected data and built the data model.

Figure 1 displays the research model. Buyer agency commission is selected as the dependent variable. This is the commission received by the real estate firm that represents the buyer. This commission is typically 2–3% of the real estate sale amount and is finally shared between the buyer agent and his/her firm. Several Facebook activities are used as independent variables, in which an average sentiment score is derived from Facebook posts.

3.1 Data Collection

We collected all the real estate transaction records from MLS (Multiple Listing Service) between January 2016 and June 2018 for the Orange county, California. We used list office phone number, list office name, and buyer agent office name to identify real estate firms on Facebook. We calculated the commission paid to the buyer agency for each sale. In order to analyze commissions by month and by year, the closing date (the date the purchase agreement was fulfilled) is included.

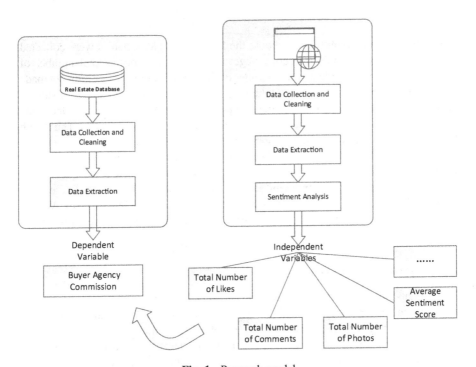

Fig. 1. Research model

Based on our datasets, there are about 400 real estate firms in Orange county. About 63% of those real estate firms have used Facebook regularly as part of their real estate marketing practice. Table 1 presents a sample list of real estate firms that have an active Facebook business page since 2016.

Table 1. Sample list of real estate agents and Facebook links

No	Real estate firm name	Facebook link
1	Allison james estates & homes	https://www.facebook.com/AJEliteHomes/#
2	Beach cities real estate	https://www.facebook.com/beachcitiesrealty/#
3	EHM real estate, inc	https://www.facebook.com/EHMRealEstate/
4	Engel & voelkers newport beach	https://www.facebook.com/EngelVolkersNewportBeach/
5	Frontier realty	https://www.facebook.com/findhomedeals/#
6	Intero real estate services	https://www.facebook.com/InteroSC/?ref=br_rs#
7	K. Hovnanian companies of California	https://www.facebook.com/khov.nocal/#
8	Pacific sterling realty	https://www.facebook.com/Pacific-Sterling-Realty-1377515665814381/
9	Pinpoint properties	https://www.facebook.com/PinpointProperties/
10	re/max prestige properties	https://www.facebook.com/RemaxPrestigePropertiesCA

To explore the opinions on Facebook, the following eight variables were collected from each real estate firm's Facebook page: total number of posts, total number of photos, total number of videos, total number of likes, total number of comments made under posts, total number of sharers, total number of links, and total number of stories. In addition, two other variables were derived to test the research hypotheses. One is the average length of posts and the other is the average sentiment score of posts. The length of a post may decide the impact of social media content on its viewers because the length shows how much effort and time that a real estate firm spent to maintain its Facebook page. Standford CoreNLP [39] was employed to calculate a sentiment score for each post. The average sentiment score was added to the Facebook dataset as a variable. The sentiment score can be used to determine the sentiment of posts. We are curious of the question: If posts sound more positive, do they tend to have more impact on sales?

3.2   Why Linear Regression?

The residuals histogram and the normal Probability Plot (PP) plot were examined to ensure that the linear regression analysis criteria were satisfied. We conducted multiple regressions. It might take months for the Facebook activities to catch up people's attention, and to finally fulfill purchase agreement. Thus, there can be a time lag between Facebook activities and a resulting real estate transaction. To account for the lagging effect, different time lags were considered in the regression analysis.

4   Results and Discussion

Multiple regressions were conducted to find out how different Facebook variables correlate with real estate sales. As explained in the prior data collection section, we collected 10 Facebook variables. Backward elimination method [40] was adopted to determine which variables should be excluded. We started with all 10 Facebook variables, then tested the deletion of each variable using criterion, which are theoretical considerations for relevance, p-values and adj. R^2 [41].

We examined the residuals histogram and the PP plot to make sure that that the linear regression analysis criteria were satisfied. After backward elimination, seven Facebook variables were included in multiple regressions. The multiple regression equation is presented as follows:

$$real\ estate\ sale = \beta_0 + \beta_1 total\ likes + \beta_2 total\ comments + \beta_3 total\ sharers + \beta_4 total\ links$$
$$+ \beta_5 total\ stories + \beta_6 post\ length + \beta_7 sentiment\ score + \varepsilon$$

$$(1)$$

The regression results from year 2016 to year 2018 are presented in Tables 2a–2c. An asterisk marks whether this relationship is significant at $p <= 0.05$.

Table 2a. Multiple regression results for year 2016

	Coefficient	Standard error
Total likes	0.163	0.182
Total comments	−0.198	0.220
Total sharers	0.097	0.107
Total links	0.338(*)	0.111
Total stories	0.170(*)	0.088
Average message length	−0.041	0.071
Average sentiment score	−0.193	0.138
R^2	0.087	
Significance F	0.008	

Table 2b. Multiple regression results for year 2017

	Coefficient	Standard error
Total likes	0.651 (*)	0.083
Total comments	−0.214	0.114
Total sharers	−0.147	0.095
Total links	0.313 (*)	0.094
Total stories	0.245 (*)	0.068
Average message length	−0.057	0.059
Average sentiment score	−0.262(*)	0.103
R^2	0.314	
Significance F	3.85E-14	

Table 2c. Multiple regression results for first six months of 2018

	Coefficient	Standard error
Total likes	0.324(*)	0.083
Total comments	−0.062	0.107
Total sharers	−0.068	0.095
Total links	0.229(*)	0.115
Total stories	0.335(*)	0.082
Average message length	−0.039	0.069
Average sentiment score	−0.200	0.139
R^2	0.170	
Significance F	1.72E-05	

As postulated by hypothesis H1, the total Facebook Likes is assumed to positively influence the real estate sales. The regression results support this hypothesis. The results in Tables 2b and 2c show the significant positive impact of total Facebook Likes on real estate sales from 2017 to 2018. Therefore, the number of Facebook Likes is positively associated with the firm's sales, probably because a higher number of Likes is a sign of the client base of this firm. The result for Facebook Likes in Table 2a (for year 2016) is not significant. One explanation is that, more recently home buyers/sellers started using Facebook Likes to help them make purchasing decisions (from 2016). Tables 2a–2c show that the sentiment is negatively associated with real estate sales, the correlation is significant in Table 2b for year 2017. Therefore, H2 is not supported. Tables 2a–2c show that the total links and total stories have strong, positive and significant correlations with real estate sales from 2016 to 2018. Thus, H3 is supported.

Time Lag. There may be a time lag between the activates on Facebook and a resulting real estate transaction, time lags were included in the following experiments and analysis. The month was used as the unit of time. First, the impact of Facebook variables on real estate sales was examined month by month. For example, Facebook variables from January 2016 were used as independent variables and the commission in January 2016 was used as dependent variable. Second, time lags were explored by choosing one-month, two-month, three-month, four-month and five-month intervals. For example, for two-month interval, if Facebook variables are from February and March 2016, then the commission in April 2016 was used as dependent variable. The regression results with and without time lags are compared in Fig. 2. The Y-axis represents a significant rate in one year. The significant rate is defined as follow.

$$\text{significant rate in one year} = \frac{\text{number of months that has significant result}}{\text{number of months in one year}} \quad (2)$$

Where a month has a significant result if the relationship in this month is significant at $p <= 0.05$.

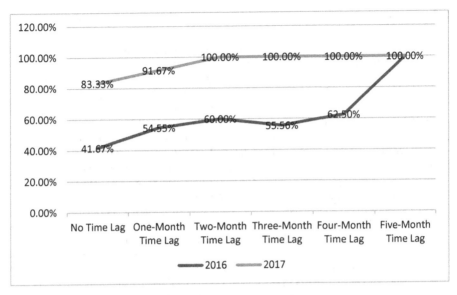

Fig. 2. Significant rates in 2016 and 2017

As shown in Fig. 2, for both 2016 and 2017, if time lag is not considered, the significant rate is the lowest. For both years, the significant rate is the highest when the time lag is at five-month. Based on the results, no matter for prediction or decision making, time lag is an important factor affecting accuracy and precision. The results in Fig. 2 support the assumption made in hypothesis H4, which is that time lag affects the impact of Facebook activities on real estate sales.

5 Conclusions and Future Research

Social media has a broad influence on businesses, and it has been used to collect customer feedback, promote brand awareness, and predict sales. In this research, we study how Facebook activities are associated with real estate sales. Using two and half years of real estate sales records for the Orange county, California and Facebook business pages of about 250 real estate firms, we find that the total numbers of Facebook Likes, links, and stories are positively associated with real estate sales; the sentiment score of Facebook posts is negatively associated with sales; when considering a time lag between Facebook activities and real estate sales, a five-month time lag leads to the highest significant rate.

As for future research, we plan to interview buyers and agents to get their further input and perception on social media's influence on real estate business. We will explore different form of social media including Twitter and YouTube and consider more attributes with a bigger real estate data set from different states to study how social media impacts real estate business.

References

1. He, W., Zha, S., Li, L.: Social media competitive analysis and text mining: a case study in the pizza industry. Int. J. Inf. Manag. **33**(3), 464–472 (2013)
2. Andzulis, J.M., Panagopoulos, N.G., Rapp, A.: A review of social media and implications for the sales process. J. Pers. Selling Sales Manag. **32**(3), 305–316 (2012)
3. Wijnhoven, F., Plant, O.: Sentiment analysis and google trends data for predicting car sales. In: 38th International Conference on Information Systems 2017 (2017)
4. Kapoor, K.K., Tamilmani, K., Rana, N.P., Patil, P., Dwivedi, Y.K., Nerur, S.: Advances in social media research: past, present and future. Inf. Syst. Frontiers **20**(3), 531–558 (2018)
5. Agarwal, A., Xie, B., Vovsha, I., Rambow, O., Passonneau, R.: Sentiment analysis of twitter data. In: Proceedings of the Workshop on Languages in Social Media, pp. 30–38. Association for Computational Linguistics, Stroudsburg, PA (2011)
6. Asur, S., Huberman, B.A.: Predicting the future with social media. In: Proceedings of the 2010 IEEE/WIC/ACM International Conference on Web Intelligence and Intelligent Agent Technology, vol. 01, pp. 492–499. IEEE Computer Society, Washington, DC (2010)
7. Gruhl, D., Guha, R., Kumar, R., Novak, J., Tomkins, A.: The predictive power of online chatter. In: Proceedings of the Eleventh ACM SIGKDD International Conference on Knowledge Discovery in Data Mining, pp. 78–87. ACM, New York (2005)
8. Bollen, J., Mao, H., Zeng, X.: Twitter mood predicts the stock market. J. Comput. Sci. **2**(1), 1–8 (2011)
9. Bartov, E., Faurel, L., Mohanram, P.S.: Can Twitter help predict firm-level earnings and stock returns? Acc. Rev. **93**(3), 25–57 (2017)
10. Joshi, M., Das, D., Gimpel, K., Smith, N.A.: Movie reviews and revenues: an experiment in text regression. In: The 2010 Annual Conference of the North American Chapter of the Association for Computational Linguistics, pp. 293–296. Association for Computational Linguistics, Stroudsburg, PA (2010)
11. Wu, L., Brynjolfsson, E.: The future of prediction: how Google searches foreshadow housing prices and sales. Economic Analysis of the Digital Economy, pp. 89–118. University of Chicago Press, Chicago (2015)
12. Schoen, H., Gayo-Avello, D., Takis Metaxas, P., Mustafaraj, E., Strohmaier, M., Gloor, P.: The power of prediction with social media. Internet Res. **23**(5), 528–543 (2013)
13. Pang, B., Lee, L.: Opinion mining and sentiment analysis. Found. Trends® Inf. Retrieval **2**((1–2)), 1–135 (2008)
14. Liu, B.: Sentiment analysis and opinion mining. Synth. Lect. Hum. Lang. Technol. **5**(1), 1–167 (2012)
15. Medhat, W., Hassan, A., Korashy, H.: Sentiment analysis algorithms and applications: a survey. Ain Shams Eng. J. **5**(4), 1093–1113 (2014)
16. Pang, B., Lee, L., Vaithyanathan, S.: Thumbs up?: sentiment classification using machine learning techniques. In: Proceedings of the ACL-02 Conference on Empirical Methods in Natural Language Processing, vol. 10, pp. 79–86. Association for Computational Linguistics, Stroudsburg, PA (2002)
17. Naik, M.V., Vasumathi, D., Siva Kumar, A.P.: An enhanced unsupervised learning approach for sentiment analysis using extraction of Tri-Co-Occurrence words phrases. In: Bhateja, V., Tavares, J.M.R.S., Rani, B.P., Prasad, V.K., Raju, K.S. (eds.) Proceedings of the Second International Conference on Computational Intelligence and Informatics. AISC, vol. 712, pp. 17–26. Springer, Singapore (2018). https://doi.org/10.1007/978-981-10-8228-3_3

18. Pak, A., Paroubek, P.: Twitter as a corpus for sentiment analysis and opinion mining. In: Proceedings of the Seventh Conference on International Language Resources and Evaluation (LREC 2010), pp. 1320–1326. European Languages Resources Association (2010)
19. Feldman, R.: Techniques and applications for sentiment analysis. Commun. ACM **56**(4), 82–89 (2013)
20. Nakov, P., Ritter, A., Rosenthal, S., Sebastiani, F., Stoyanov, V.: SemEval-2016 task 4: sentiment analysis in Twitter. In: Proceedings of the 10th International Workshop on Semantic Evaluation, pp. 1–18, Association for Computational Linguistics (2016)
21. Severyn, A., Moschitti, A.: Twitter sentiment analysis with deep convolutional neural networks. In: Proceedings of the 38th International ACM SIGIR Conference on Research and Development in Information Retrieval, pp. 959–962. ACM, New York (2015)
22. Saif, H., He, Y., Fernandez, M., Alani, H.: Contextual semantics for sentiment analysis of Twitter. Inf. Proc. Manag. **52**(1), 5–19 (2016)
23. Pandey, A.C., Rajpoot, D.S., Saraswat, M.: Twitter sentiment analysis using hybrid cuckoo search method. Inf. Proc. Manag. **53**(4), 764–779 (2017)
24. Ortigosa, A., Martín, J.M., Carro, R.M.: Sentiment analysis in Facebook and its application to e-learning. Comput. Hum. Behav. **31**, 527–541 (2014)
25. Meire, M., Ballings, M., Van den Poel, D.: The added value of social media data in B2B customer acquisition systems: A real-life experiment. Decis. Support Syst. **104**, 26–37 (2017)
26. Cambria, E.: Affective computing and sentiment analysis. IEEE Intell. Syst. **31**(2), 102–107 (2016)
27. Poecze, F., Ebster, C., Strauss, C.: Social media metrics and sentiment analysis to evaluate the effectiveness of social media posts. Procedia Comput. Sci. **130**((C)), 660–666 (2018)
28. Nguyen, T.H., Shirai, K., Velcin, J.: Sentiment analysis on social media for stock movement prediction. Expert Syst. Appl. **42**(24), 9603–9611 (2015)
29. Fang, X., Zhan, J.: Sentiment analysis using product review data. J. Big Data **2**(1), 5 (2015)
30. Boldt, L.C., et al.: Forecasting nike's sales using Facebook data. In: 2016 IEEE International Conference on Big Data, pp. 2447–2456. IEEE (2016)
31. Lee, K., Lee, B., Oh, W.: Thumbs up, sales up? the contingent effect of Facebook likes on sales performance in social commerce. J. Manag. Inf. Syst. **32**(4), 109–143 (2015)
32. Mazzucchelli, A., Chierici, R., Ceruti, F., Chiacchierini, C., Godey, B., Pederzoli, D.: Affecting brand loyalty intention: the effects of UGC and shopping searches via Facebook. J. Glob. Fashion Mark. **9**(3), 270–286 (2018)
33. Geetha, M., Singha, P., Sinha, S.: Relationship between customer sentiment and online customer ratings for hotels-an empirical analysis. Tourism Manag. **61**, 43–54 (2017)
34. Dini, L., Bittar, A., Robin, C., Segond, F., Montaner, M.: SOMA: The Smart Social Customer Relationship Management Tool: Handling Semantic Variability of Emotion Analysis With Hybrid Technologies. Sentiment Analysis in Social Networks, pp. 197–209 (2017)
35. He, W., Chen, Y.: Using blog mining as an analytical method to study the use of social media by small businesses. J. Inf. Technol. Case Appl. Res. **16**(2), 91–104 (2014)
36. Pedhazur, E.J., Kerlinger, F.N.: Multiple Regression in Behavioral Research. Holt, Rinehart and Winston, New York (1973)
37. Bing, L., Chan, K.C.C., Ou, C.: Public sentiment analysis in Twitter data for prediction of a company's stock price movements. In: 2014 IEEE 11th International Conference on E-Business Engineering, pp. 232–239. IEEE (2014)
38. Kotler, P.J.: Marketing Management: Analysis, Planning, Implementation, and Control, 8th edn. Prentice Hall, Englewood Cliffs (1994)

39. Manning, C., Surdeanu, M., Bauer, J., Finkel, J., Bethard, S., McClosky, D.: The stanford coreNLP natural language processing toolkit. In: Proceedings of 52nd Annual Meeting of the Association for Computational Linguistics: System Demonstrations, pp. 55–60. Association for Computational Linguistics, Stroudsburg, PA (2014)
40. Myers, R.H., Myers, R.H.: Classical and Modern Regression With Applications, vol. 2. Duxbury Press, Belmont (1990)
41. Studenmund, A.H.: Using Econometrics: A Practical Guide (5 th). Pearson Education Inc, Boston (2006)

An Ecological Business Model for Intelligent Operation and Maintenance of Urban Infrastructure

Juan Du[1,2(✉)], Xin Wang[1,3], and Vijayan Sugumaran[4,5]

[1] SHU-UTS SILC Business School, Shanghai University, Shanghai, China
ritadu@shu.edu.cn
[2] School of Building Construction, College of Design,
Georgia Institute of Technology, Atlanta, USA
[3] SHU-SUCG Research Centre for Building Industrialization,
Shanghai University, Shanghai, China
[4] School of Business Administration, Oakland University, Rochester, MI, USA
[5] Center for Data Science and Big Data Analytics, Oakland University,
Rochester, MI, USA

Abstract. With the increasing number of urban infrastructure construction projects under the PPP (Public-Private-Partnership) investment, the traditional operation and maintenance model presents a number of problems, such as low efficiency and high cost of operation & maintenance, fragmentation of stakeholders, difficulty in estimating the value of facility assets and utilizing the isolated multi-source data. In view of this, this research proposes a new model for Intelligent Operation and Maintenance of Urban Infrastructure (IOMUI), which is based on the exploration of the claims from various stakeholders under the PPP model, and combined with emerging ICT technologies. Through the analysis of the current facility industry chain, our research proposes a shared ecosystem, which extends the vertical industrial chain, and ultimately forms the model of IOMUI. The proposed ecological business model provides a reference for the reform and upgrading of the OMUI industry, and also contributes to the sustainable development of the industry.

Keywords: PPP (Public-Private-Partnership) · Urban infrastructure ·
Intelligent operation and maintenance · Business model · Ecosystem

1 Introduction

With the continuous promotion and development of PPP (Public-Private-Partnership) projects, the operation and maintenance management of urban infrastructure (OMUI) has undergone great changes. The PPP investment model means that the public government department signs a franchising contract with the investment corporation in the form of government procurement. And the investment corporation is responsible for fundraising and construction, as well as operation and maintenance of infrastructure during the government licensed franchise period. (The investment corporation is a limited liability firm consisting of bidding construction firm, a service management

© Springer Nature Switzerland AG 2019
J. J. Xu et al. (Eds.): WEB 2018, LNBIP 357, pp. 15–25, 2019.
https://doi.org/10.1007/978-3-030-22784-5_2

firm or a third party that invests in the project). When the franchise expires, the government is in charge of evaluating the asset of the urban infrastructure project. If the assessment is passed, the project will be transferred to the government for management. Therefore, PPP is a modern financing model that is based on the cooperation concept of "win-win" or "multi-win". Substantially, adopting this form of financing, the government can achieve effective construction and operation by granting the private company long-term franchise. While the investment corporation obtains revenue by effectively managing and providing services for infrastructure projects during the franchise period. However, during the project implementation process, there are many defects and shortcomings in the traditional model of OMUI. For example, the investors can hardly know the condition, remaining life and fixed asset value of the urban infrastructure. It will result in an intangible increase of investment risk and investors may lose interest in follow-on investments. Besides, due to the current corrective maintenance model, the maintenance companies are unable to predict in advance and maintain timely, resulting in a sharp increase in operating costs in the middle and late stages. In essence, a new model that matches the OMUI stage is needed and the reform of OMUI is imperative.

In previous studies, there have been many researches that integrate advanced ICT technologies into the information networks or information system architecture of smart city, such as Internet of Things, cloud computing, mobile service, data mining and wireless technology [3, 10, 11, 15, 21, 23]. However, the researches establishing open innovation model based on network system architecture are lacked. Hans Schaffers et al. [22] proposed User Co-creation Innovation Ecosystem for smart citiy. But this research focused on the feasibility of the platform, without puting forward the user's value proposition, as well as the core modules in the platform. In addition, there are also studies on the establishment of urban infrastructure platforms [17, 18]. But these studies are mainly carried out the information system architecture to support decision making and early warning, without forming an open ecological platform around the multilateral stakeholders and users of urban infrastructure.

Therefore, in order to form an open ecological model that matches the OMUI under PPP, this research combines with the new generation of information technology to explore the value proposition of stakeholders, and also innovatively integrates the characteristics of ecological platform into the new model, which will be of great significance for stakeholders of PPP projects and the development of the industry.

2   Industrial Background and Business Model

2.1   Industrial Background

Constrained by information technology and management model, the traditional model of OMUI has many deficiencies, mainly including the following:

Lack of the Evaluation Mechanism for Infrastructure Assets. The investment corporation invests massive funds in the early stage, but could hardly know the exact condition, remaining life and asset value of urban infrastructure that comes into service. Usually, decisions can only be made based on financial analysis and risk prediction data that is superficial and estimated, due to lack of sufficient decision data. Currently, the traditional operation and maintenance model cannot monitor the real-time information of the facility, and there is no effective evaluation method for the performance of operation and maintenance. Therefore, it can be difficult to know the exact service performance of the facility and evaluate the asset value. At present, a large part of infrastructure is seriously damaged before reaching the service life, resulting in huge waste of maintenance funds. Besides, the calculation of maintenance costs is unreasonable due to the low transparency of capital inflow, giving rise to increased operation and maintenance costs in the later stage.

Lack of the Intelligent Service to the Infrastructure Sustainable Management. With the continuous development of modern cities, the needs of urban residents for infrastructure are more diversified. The existing operation and maintenance model is mainly focused on corrective maintenance and emergency management, instead of predicting, warning and dredging in advance. For example, during the peak travel period, the increase of uncertainties will lead to large traffic congestion and many other management problems.

Lack of Good Facilities Inspection & Detection Methods. The facility maintenance technician usually adopts manual means, such as daily patrol, to collect abnormal data, input information and so on. However, this method cannot capture the abnormal information of facilities or collect accurate data timely. Besides, it aggravates the shortage of human resources for facility maintenance.

With increasing service time of urban infrastructure, the probability of occurrence of the abnormal situation is growing continuously. The existing technology is unable to accurately determine causes of abnormality or predict occurrence. The backward technology and lack of equipment results in missing the best maintenance time and affecting life of facilities, further increasing the cost.

At present, in spite of regular data collection and management for some urban infrastructure, it still stays in the information processing stage. The data mining and analysis is lacked and data utilization rate is low. Most urban infrastructure data exists independently in related companies, so information querying is very inconvenient. Therefore, the degree of information sharing between each other is low, and the maintenance information channels between industries are not smooth.

Lack of Scientific Decision-making Tools. The maintenance information channels between related departments and industries are not smooth and the database is incomplete. And due to the lack of scientific tools, decision-making mainly depends on the judgment of experienced workers. The decision information needs to be conveyed layer by layer, which is inefficient and prone to errors. Without reasonable method of project evaluation, the establishment of maintenance plan is unreasonable and unmatched with the actual maintenance requirements.

The malpractice of management and technology has been a major barrier to the development of OMUI. Therefore, it is of great importance to increase technical investment. Besides, exploring the reasonable model for OMUI is necessary, which can match with the continuous reform of maintenance technology.

2.2 Business Model

To develop a good business model, scholars have conducted multi-level research based on different fields and perspectives. However, there is still no exact definition, forming a "concept jungle" of business model. Table 1 summarizes the classic definitions of business model.

Table 1. The definitions of Business Model

Authors (Year)	Definitions
Linder and Cantrell (2000) [13]	Business model is a logical set of value creation, and the transaction, cash inflow and other processes of an organization are carried out around this logical set (Linder and Cantrell 2000)
Chesbrough and Rosenbloom (2002) [4]	Business model is a network structure composed of a company and its partners, which creates and delivers value to the target customers and gain profits (Chesbrough and Rosenbloom 2002)
Osterwalder and Pigbeur Tucci (2005) [16]	Business model refers to the value provided by the company for customers and how the value is created, marketed and transmitted between the company's internal networks to generate profits (Osterwalder and Pigbeur Tucci 2005)

In order to express the connotation and category of business model more clearly, it is necessary to present the components of business model. Currently, the understanding of the components of business model can be summarized in Table 2.

Table 2. Business model elements of different points of view

No	Source	Components
1	Horowitz (1996) [9]	Price, product, distribution, organizational characteristics, technology
2	Lambert (2008) [12]	Value, customer, return, channel, investment, partner, appreciation
3	Afuah (2002) [1]	Marketing field, customer value, related behaviors, implementation efficiency, sustainability, pricing strategy, revenue model
4	Markides (1999) [14]	Product innovation, customer, infrastructure management, financial status
5	Dirk (2009) [6]	Enterprise operation, customer, marketing, network internality, network externality
6	Richardson (2008) [19]	Value proposition, value delivery, value acquisition
7	Wirtz (2011) [20]	Product, user interface, infrastructure, finance

(continued)

Table 2. (*continued*)

No	Source	Components
8	Chesbrough and Rosenbaum (2002) [4]	Product positioning, target customers, internal value chain, external value network, property status, bidding strategy
9	Linder et al. (2000) [13]	Organizational structure, interconnection, market positioning, channel, process, pricing, revenue model
10	Dubosson-Torbay (2002) [7]	Product services, customers, partners, channels and revenue status
11	Applegate (2001) [2]	Concepts, capabilities and values
12	Hamel (2001) [8]	Core strategy, strategic resources, value network, customer interface.
13	Demil and Lecocq (2010) [5]	Resources and capabilities, organizational structure, value delivery
14	Osterwalder et al. (2005) [16]	Value proposition, channel, customer, market positioning, main business, core capability, ally partner, cost composition, profit model

Based on the components of business model above, this research will analyze the elements of the business model for IOMUI and provide further ecological trend development to IOMUI.

3 The Business Model of IOMUI

3.1 Value Propositions of the Stakeholders

In order to better design the new model of OMUI which is based on multilateral platform, and better serve the project investment corporation, it is necessary to accurately understand the demand of the project investment corporation for the OMUI under PPP. The value proposition mining analysis table of stakeholders is shown in Table 3.

Table 3. The analysis of value propositions for stakeholders of OMUI

Stakeholders	Analysis of value propositions
Government/Investors	1. Excellent data analysis methods 2. Accurate evaluation of the value of facilities 3. Master facilities performance and fixed assets status 4. Operation and maintenance cost statistics 5. Risk assessment of the project
Operating companies	1. Optimize facilities operation and reduce operating costs 2. Visual management tools 3. Data analysis service

(continued)

Table 3. (*continued*)

Stakeholders	Analysis of value propositions
Maintenance companies	1. Precise prediction of abnormal data 2. Accurate means of equipment detection 3. Efficient maintenance work and professional technology
Suppliers of equipment/materials	1. Obtain timely market demand information 2. Accurate data information
Third party provider of software development/IT service	1. Master market demand 2. Implement technology and service output to solve customer data application problems
Urban residents	1. Safe and reliable facilities for convenient and efficient trip 2. Intelligent facilities information service 3. Timely repair of the damaged facilities

3.2   Intelligent Trends for OMUI

According to the value proposition of participating parties, this research proposes an ecosystem-based model of IOMUI. The so-called intelligent model is under the support of new technology (e.g. cloud computing technology, data mining, Internet of Things) to establish a multilateral platform for OMUI. It is aimed at providing services of intelligent awareness, prediction, evaluation and transaction for stakeholders, and forming a highly efficient, safe and economic model of OMUI. So the intelligent model includes the following aspects:

Intelligent Awareness. The function of urban infrastructure monitoring and detection. During the construction and production of infrastructure, intelligent sensors will be implanted to acquire real-time information during the subsequent stage of OMUI. And the real-time data of infrastructure structure, equipment, environment and operation & maintenance process are also collected to monitor the occurrence of abnormal data.

Intelligent Prediction. The function of forecasting possible abnormal situation of facilities. Through the collection of all kinds of abnormal data and big data mining, the health condition of all kinds of facilities can be understood and predicted periodically to some extent, so as to realize intelligent prediction and guarantee maintenance work before occurrence of malfunction.

Intelligent Transaction. Remote manipulation or query function of large-scale cash flow that generated in project through the platform. All kinds of urban infrastructure projects in the platform need capital inflows, including massive operation and maintenance funds. So this function provides transparency of funds, and the authorized stakeholders can check the capital generated in each process. The project transaction includes project tendering & bidding, and the cash flow of construction, operation & maintenance are manipulated in this function module.

Intelligence Evaluation. Through data analysis and modeling of the infrastructure environment sensor data, feasibility evaluation of each project can be achieved. Other services such as safety evaluation, risk assessment for all kinds of urban infrastructure can be completed in the module.

The value proposition of IOMUI is based on intelligent data analysis, and provides users with data analysis of urban infrastructure project in each stage through the cloud platform, in order to optimize the OMUI. For urban infrastructure, the intelligent model will provide the real-time service status of infrastructure, predict the occurrence of malfunction, and maintain facilities prone to abnormal conditions in advance. Besides, the supervision of operation and maintenance process and quality verification can be achieved. Otherwise, it can provide the facilities assessment of various indicators to ensure the feasibility of all stages and reduce project risks. Meanwhile, it will provide users with a safe and reliable fund trading platform to ensure the transparency and safety of cash flow in the project bidding, construction, operation and maintenance stage.

3.3   The Design of Business Model of IOMUI

According to the literature review of business model, there are many complicated elements of business model. To ensure that every factor in the business model is analyzed thoroughly, this research selects the most classic elements of Alexander Osterwalder business model to design business model for IOMUI. The structure diagram is shown in Fig. 1. Combined with the 9 elements theory of business model components, the business model of IOMUI is shown as Fig. 2.

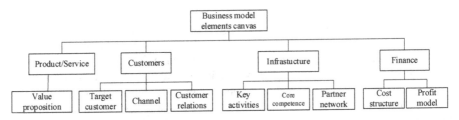

Fig. 1. The structure chart of business model elements

Partners	Key activities	Value proposition	Customer relations	Target customers
Government regulators, Outsourcing company, maintenance company, Trade associations, Internet technology companies, Platform operating company, Platform sponsor	Facilities operation management, Facilities timely maintenance, Facility anomaly forecast, Project investment transaction, Project index evaluation, Multilateral platform operation, Data analysis and mining, Service and technical consultation.	Intelligence perception, Intelligence prediction, Intelligence transaction, Intelligence evaluation A new model of OMUI with high efficiency, safety and economy has been formed	Service relationship, Customer support, Partner relationship	Government, Investment companies, Operation companies, Maintenance companies, Monitoring companies, Equipment/ material suppliers, IT service outsourcing supplier
	Key abilities: Database of IOMUI, Multilateral platform, Big data industry		**Channels** Online trading platform, Real-time operating platform, Sharing platform, Other multilateral platforms	

Cost structure:	Revenue stream
Platform O&M cost, Human resources cost, Facility maintenance cost, Emergency maintenance cost	Return on project investment, Subsidies, Facility charges, Platform data transaction, Consultancy charges, Advertisement fees, Platform lease fees, Membership service fee

Fig. 2. The intelligent operation and maintenance business model canvas

4   The Ecosystem of IOMUI

4.1   The Ecological Chain of IOMUI

At present, the OMUI mainly involves the government, operating companies, maintenance companies, materials & equipment suppliers, testing companies and users. The industrial chain is mainly a linear chain of "planning & design—construction—operation & maintenance". The government invests in the construction of various infrastructure, and entrusts the testing companies to carry out the detection. And the operation and maintenance companies are in charge of the operation and maintenance. While the suppliers provide relevant equipment and material during the construction, and provides services for the users to obtain returns.

By building the platform of IOMUI, many vertical expansion chains can be formed. For example, the chain of "infrastructure—value assessment & consultation" is established through infrastructure value assessment; the chain of "infrastructure—intelligent sensor company" chain is based on the collection of infrastructure data; the "infrastructure—IT service outsourcing company" chain is built through the establishment of the infrastructure operation and maintenance multilateral platform. The design of overall intelligent operation and maintenance industrial chain is shown in Fig. 3 below.

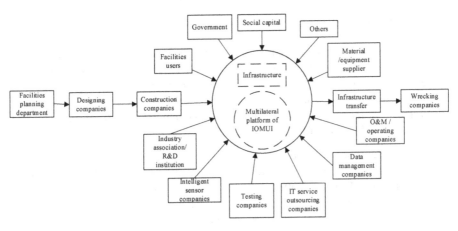

Fig. 3. Model diagram of ecological chain of IOMUI

4.2 The Ecosystem of IOMUI

When the vertical industrial chain of intelligent operation and maintenance continues to extend, different organizations and groups will gather, interact and cooperate with each other, and finally form a shared ecosystem of intelligent operation and maintenance model, shown as Fig. 4 below.

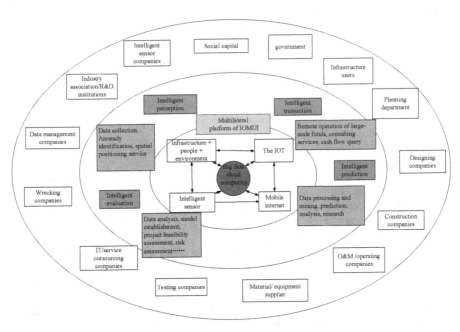

Fig. 4. The model of shared ecosystem of intelligent operation and maintenance

5 Conclusion

Based on the analysis of the current situation of OMUI, this research fully mines the value proposition of the urban infrastructure stakeholders and operation & maintenance requirements, which is combined with the characteristics of a new generation of information technology. Then the research proposes the corresponding solution, namely the new model of IOMUI, and the ecosystem framework of IOMUI. The research work includes the following:

(1) Through the analysis of the existing technical capacity of information technology, this research proposes the methods to collect, analyze and utilize the data of urban infrastructure. The data platform is built to cope with the shortages of traditional OMUI. And the related concepts and value propositions of IOMUI are formed.
(2) Based on the retrieval of literatures and the value proposition, the basic model of IOMUI is formed. Combined with basic theory of business model, a new model of IOMUI is designed.
(3) Through the analysis of the industrial chain of OMUI, a multilateral platform shared ecosystem is formed, which extends from the single industrial chain to the vertical industrial chain and finally forms the model of IOMUI.

The ecological business model proposed in this research provides a reference for the reform and upgrading of the UIOM industry, and also contributes to the sustainable development of the industry.

Acknowledgment. This work is supported in part by the National Natural Science Foundation of China under Grant 71701121, and the Chinese Ministry of Education of Humanities and Social Science Project under Grant 17YJC630021.

References

1. Afuah, A., Tucci, C.L.: Internet Business Models and Strategies: Text and Cases. McGraw-Hill Higher Education, New York (2002)
2. Applegate, L.M.: Emerging e-Business Models: Lessons from the Field. Harvard Business School, Boston (2001)
3. Crang, M., Graham, S.: Sentient cities ambient intelligence and the politics of urban space. Inf. Commun. Soc. 10(6), 789–817 (2007)
4. Chesbrough, H., Rosenbloom, R.S.: The role of the business model in capturing value from innovation: evidence from xerox corporation's technology spin-off companies. Ind. Corp. Change 11(3), 529–555 (2002)
5. Demil, B., Lecocq, X.: Business model evolution: in search of dynamic consistency. Long Range Plan. 43(2–3), 227–246 (2010)
6. Dirk, D., Danny, L.: Plausibility in early stages of architectural design: a new tool for high-rise residential buildings. Tsinghua Sci. Technol. 14(3), 327–332 (2009)
7. Dubosson-Torbay, M., Osterwalder, A., Pigneur, Y.: E-business model design, classification, and measurements. Thunderbird Int. Bus. Rev. 44(1), 5–23 (2002)
8. Hamel, G.: Leading the revolution. Strategy Leadersh. 18(1), 212–213 (2001)

9. Horowitz, A.S.: The real value of VARS: resellers lead a movement to a new service and support. Mark Comput. **16**(4), 31–36 (1996)
10. Jin, J., Gubbi, J., Marusic, S., Palaniswami, M.: An information framework for creating a smart city through internet of things. IEEE Internet Things J. **1**(2), 112–121 (2014)
11. Kitchin, R.: The real-time city? big data and smart urbanism. GeoJournal **79**(1), 1–14 (2014)
12. Lambert, S.: A conceptual framework for business model research. In: 21st Bled Conference Collaboration: Overcaming Boundaries Through Multi-Channel Interaction, 15–18 June 2008 (2008)
13. Linder, J.C., Cantrell, S.: Changing Business Models: Surveying the Landscape. Accenture Institute for Strategic Change (2000)
14. Markides, C.C.: A dynamic view of strategy. Sloan Manag. Rev. **40**(3), 55–63 (1999)
15. Mohanty, S.P., Choppali, U., Kougianos, E.: Everything you wanted to know about smart cities. IEEE Consum. Electron. Mag. **5**(3), 60–70 (2016)
16. Osterwalder, A., Pigneur, Y., Tucci, C.L.: Clarifying business models: origins, present, and future of the concept. Commun. Assoc. Inf. Syst. **16**(16), 751–775 (2005)
17. Quintero, A.: Prototyping an intelligent decision support system for improving urban infrastructures management. Eur. J. Oper. Res. **162**(3), 654–672 (2005)
18. Quintero, A., Konaré, D., Pierre, S.: Using mobile agents for managing control and alarms in urban infrastructures. In: Pierre, S., Glitho, R. (eds.) MATA 2001. LNCS, vol. 2164, pp. 195–209. Springer, Heidelberg (2001). https://doi.org/10.1007/3-540-44651-6_19
19. Richardson, J.: The business model: an integrative framework for strategy execution. Strateg. Change **17**(5–6), 133–144 (2008)
20. Wirtz, B.: Business Model Management: Design Process Instruments. Gabler University of Administrative Science, Speyer (2011)
21. Sanchez, L., Muñoz, L., Galache, J.A., Sotres, P., Santana, J.R., Gutierrez, V., et al.: Smartsantander: iot experimentation over a smart city testbed. Comput. Networks **61**(6), 217–238 (2014)
22. Schaffers, H., Komninos, N., Pallot, M., Trousse, B., Nilsson, M., Oliveira, A.: Smart Cities and the Future Internet: Towards Cooperation Frameworks for Open Innovation. In: Domingue, J., Galis, A., Gavras, A., Zahariadis, T., Lambert, D., Cleary, F., Daras, P., Krco, S., Müller, H., Li, M.-S., Schaffers, H., Lotz, V., Alvarez, F., Stiller, B., Karnouskos, S., Avessta, S., Nilsson, M. (eds.) FIA 2011. LNCS, vol. 6656, pp. 431–446. Springer, Heidelberg (2011). https://doi.org/10.1007/978-3-642-20898-0_31
23. Yovanof, G.S., Hazapis, G.N.: An architectural framework and enabling wireless technologies for digital cities & intelligent urban environments. Wireless Pers. Commun. **49**(3), 445–463 (2009)

Data-Driven Business Models and Professional Services Firms: A Strategic Framework and Transitionary Pathways

Erwin Fielt[✉], Kevin C. Desouza, Guy Gable, and Peter Westerveld

Queensland University of Technology, Brisbane, Australia
{e.fielt, kevin.desouza, g.gable,
peter.westerveld}@qut.edu.au

Abstract. Many organizations and industries are undergoing a significant transformation due to digital technologies. In our research, we study digital business model innovation in relation to Professional Services Firms (PSFs). In this conceptual paper, we contrast the traditional, human-centered, knowledge-intensive business model of PSFs with the new, computer-centered, data-driven business model that is developing due to the rise of big data, advanced data analytics and artificial intelligence. To better understand if, when and how data-driven business models may disrupt PSFs, we provide a strategic framework for identifying and analyzing the options for PSFs in relation to the nature and scope of their value proposition. We suggest several possible transitionary pathways using digital technology for augmentation or automation and the need so scale across services and industries. As such this paper provides valuable insights to academics and practitioners into how PSFs might develop new business models given the nature of their service offerings and industry positions.

Keywords: Digital innovation · Digital transformation · Knowledge-intensive business models · Data-driven business models · Professional Services Firms · Strategic innovation

1 Introduction

The ongoing proliferation of digital technologies, such as mobile technology, cloud computing, data analytics, and internet of things, is changing the way people live their lives, transforming the way organizations conduct their business, and creating new kinds of services and products. The World Economic Forum states that '*the future of countries, businesses, and individuals will depend more than ever on whether they embrace digital technologies*' (Baller et al. 2016, p. v). These new digital technologies, and the disruptive innovations they enable and drive, often require novel, digital business models (Fichman et al. 2014; Fielt and Gregor 2016).

Business models matter; the same idea or technology taken to market through two different business models will yield two different economic outcomes (Chesbrough 2010). A business model describes how an organization creates and captures customer value by addressing the customer, value proposition, relationships & channels, resources, activities & partners, and revenues & costs (Osterwalder and Pigneur 2010). Value creation for customers is at the core of business models (Fielt 2013; Massa et al. 2017) and the value proposition is seen as the core component (Demil et al. 2015; Morris et al. 2005). Business model innovation is targeting new ways for organizations to create and capture customer value (Chesbrough 2006) and has become a prominent theme (e.g., Foss and Saebi 2017; Massa and Tucci 2014). However, digital business model innovation and its relationship with digital transformation is not well understood (DaSilva et al. 2013). The few studies conducted (e.g., Westerman et al. 2014) give little attention to specific challenges of business model innovation, let alone contextualise these understandings to particular industries and address the strategic options within those industries.

We examine digital business model innovation in relation to Professional Services Firms (PSFs). PSFs are knowledge-intensive, which is expected to engender specific issues in relation to how data analytics and machine learning innovations will impact their strategic and operational postures. We contrast the traditional, knowledge-intensive business model of PSFs with the new, data-driven business model that is emerging due to the rise of big data, advanced data analytics and artificial intelligence. We see these two business models as archetypes (Greenwood and Hinings 1993), with the former emphasizing the traditional human-centered approach of PSFs and the latter emphasizing a potentially disruptive, computer-centered approach. Note that we use these two extreme images of business models for analytical purposes, as reality is more complicated and messier and many other business models will be possible, including hybrid forms of these two models.

Moreover, data-driven business models for PSFs may differ from models in other industries due to unique PSF characteristics, and the consequent unique roles data and technology may play. In some professional services industries there is data which is a shared resource which can be exploited by new technologies e.g. in the legal services industry, models have been developed that predict the behaviour of the Supreme Court of the United States, based on historical Supreme Court justice votes and generally available case data (Katz et al. 2017). Further, technologies such as machine learning are making it possible for digital agents to complete legal forms and even provide recommendations for simple queries and tasks.

Our research objective is to better understand if, when and how data-driven business models may disrupt the knowledge-intensive business models of PSFs. We provide a framework for identifying and analyzing the strategic options in relation to the nature and scope of their value proposition. Existing PSFs are not homogenous; their services and strategies vary. We anticipate, as a consequence, the issues faced by PSFs also vary, as will alternative ways forward in the face of change. We follow the established idea of service research that we need to turn to classification schemes to more fully address the complexities of services (Cook et al. 1999). Lovelock (1983), for example, has classified services to gain strategic insights. We have developed a framework that classifies PSFs based on nature of their value proposition in terms of the level of customization of their service offering and the technology intensity of the services they provide. It also classifies PSFs based on the scope of their value proposition in terms of the range and degree of focus of the services they offer. Based on an analysis of these classifications we suggest several possible transitionary pathways.

The remainder of the paper is organized as follows. First, we take a closer look at PSFs and discuss their traditional, human-centered, knowledge-intensive business models. Next, we address digital innovation, in particular, new, computer-centered, data-driven business models. Then we address their impact on PSFs by introducing a strategic framework and discuss possible transitionary pathways. Finally, we end with concluding remarks, limitations and future research.

2   Professional Services Firms and Traditional Knowledge-Intensive Business Models

The professional services industry has emerged as one of the most rapidly growing, profitable, and significant sectors of the global economy (Empson et al. 2015). PSFs are knowledge intensive organizations that facilitate economic and commercial exchange by providing advice to business (Greenwood et al. 2006). They are comprised primarily of professionals and their key resources are intellectual capital and expertise. Løwendahl et al. (2001) note that 'PSFs … employ a very high percentage of highly educated people, and they are extremely dependent on their ability to attract, mobilize, develop and transform the knowledge of these employees to create value for their clients.' Empson et al. (2015) state that PSFs are defined by four characteristics they all possess, to varying degrees: (1) the primary activity of applying specialist knowledge to create customized solutions to clients' problems, (2) specialist technical knowledge of professionals and in-depth knowledge of their clients as core assets, (3) governance arrangements with extensive individual autonomy and contingent managerial authority, where core producers own or control core assets, and (4) an identity where core producers recognize each other as professionals and are recognized as such by clients and competitors. Maister (1993) notes that PSFs perform three types of work: (1) procedural work for which the solution/approach is (mostly) well-known and the focus is on efficiency, (2) grey hair work requiring skills and experience, and (3) brain work requiring expertise and innovation. He also states that there are two important characteristics of professional work: (1) professional services involve a high degree of customization in their work and (2) professional services have a strong component of face-to-face interaction with the client.

With increasing availability and access to data for anyone ('data democratization') and new ways of creating and leveraging knowledge (e.g., crowdsourcing), the strategic position of PSFs as gatekeepers of knowledge is potentially under threat. For example, in the legal services industry, established firms are confronted with new start-ups introducing new digital services such as legal decision predictions, e.g., Case Crunch (see www.case-crunch.com) and 'robot lawyers,' e.g., DoNotPay (see www.donotpay.com). Increasingly automated, data-driven services that form part of, or affect the processes and delivery of the services and products PSF's provide, could also impact their capabilities, resources and partners. For example, legal firms are looking into the application of smart contracts and risk management services utilizing blockchain (Shah 2018). This demands specific capabilities and digitally integrated business partner/client networks that are likely to require business model changes by legal PSFs. We next look in more detail at these data-driven business model that are emerging, often being pioneered by tech firms.

3   Digital Innovation and New Data-Driven Business Models

The ongoing proliferation of digital technologies, such as mobile technology, social media, cloud computing, and internet of things, is changing the way people live their lives, transforming the way organizations conduct their business, and creating new kinds of services and products. Brynjolfsson and McAfee (2014) refer to the 'Second Machine Age', where technological progress is exponential (Moore's law), digital and combinatorial. Digital innovation refers broadly to digitization of innovation processes and outcomes with digital technologies playing the role of enabler or trigger (Nambisan 2013; Nambisan et al. 2017). Digitization refers to the conversion from analogue and physical to digital as a necessity for digital innovation (Fichman et al. 2014; Yoo et al. 2010). Yoo et al. (2010) view digitization as 'the encoding of analog information into digital format' (p. 725) with a focus on embedding digital capabilities into physical products. Fichman et al. (2014) refer to 'the practice of taking processes, content or objects that used to be primarily (or entirely) physical or analog and transforming them to be primarily (or entirely) digital' (p. 333). Fichman et al. (2014) suggest three types of digital innovation: product, process and business model innovation. Although these types of innovation are distinct, they are often closely linked in digital innovations and the lines between them can get blurred. Fichman et al. define business model innovation as 'a significantly new way of creating and capturing business value that is embodied in or enabled by IT' (p. 335).

With the rise of big data, advanced data analytics and artificial intelligence (Chen et al. 2012; Günther et al. 2017; Newell and Marabelli 2015), new business models are becoming data-driven (Hartmann et al. 2016). Data-driven business models are shaped by critical data-driven elements. Data is the key resource, the process of turning data into value the key activity, data-enriched or data-driven products and services the value proposition, and monetized data the revenue stream (Hartmann et al. 2016; Schüritz et al. 2017; Wixom and Ross 2017). Hartmann et al. (2016) in an empirical analysis of start-ups, identified six types of data-driven business model differentiated by the key data source and key data activity. Firms often have multiple options when it comes to creating data-driven business models. For example, the Climate Corporation initially used data analytics to provide weather insurance to agricultural firms, then moved out of weather insurance and became a digital agriculture platform that assists agricultural firms to make data-driven decisions that maximize their returns.

Firms can use data, analytics and algorithms to improve and/or innovate their business model (Günther et al. 2017; Loebbecke and Picot 2015; Woerner and Wixom 2015). To *improve* their business model, firms can use data, analytics and algorithms to refine and optimize their business processes and decision making (Woerner and Wixom 2015). To *innovate* their business model, firms can use data, analytics and algorithms to find new ways of generating revenues, enter new markets, and even explore new sources of competitive advantage through strategic renewal via data monetization and digital transformation (Woerner and Wixom 2015). According to a study by McKinsey (Chin et al. 2017), advanced analytics can create new opportunities and disrupt entire industries. For example, some companies now charge for their analytics-enabled service rather than directly selling the product. Günther et al. (2017) note that improvement and innovation approaches can be mixed and may even happen in sequence. Loebbecke and Picot (2015) warn that traditional firms may fail to benefit from big data analytics as, on the one hand, the 'improvement' of business models may not be enough for lasting competitive advantage due to the commoditization of big data solutions and, on the other hand, the 'innovation' of business models may be a struggle for these firms when established business models are disrupted.

4   A Strategic Framework for PSFs

To better understand if, when and how data-driven business models may disrupt the traditional knowledge-intensive business models of PSFs, we provide an initial framework for identifying and analyzing the strategic options of PSFs in relation to the nature and scope of their value proposition (Fig. 1). Maister (1993) has stated that professional work has two important characteristics (degree of customization and face-to-face interaction) and PSFs performing three types of work (procedural, grey hair and brain work). Based on these ideas, PSFs can be classified based on the level of customization and technology intensity of the service/s they provide, which relates directly to the value proposition as core component of the business model and as such will also influence the other components (matrix I in Fig. 1).

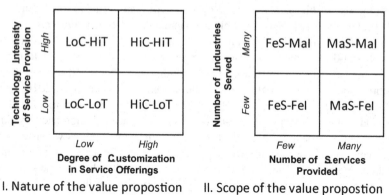

Fig. 1. Strategic framework for professional services.

A LoC-LoT PSF is one that provides generic services and does not rely a lot on technology as a key differentiator (e.g. traditional real-estate agents). A HiC-LoT PSF that offers services with a high degree of customization and low technology intensity is exemplified by boutique firms that work with a small selection of clients such as luxury real-estate agents. A LoC-HiT PSF that is low on customization but high on technology intensity would look like a traditional management consulting firm. A HiC-HiT PSF that is high on both dimensions would look like a high-end architecture or engineering services firm.

LoC-LoT PSFs will face severe pressures due to digital disruptions. The disruptions will come by way of technologies that automate service offerings while at the same time enabling customization and personalization of the delivery of the service. These disruptions will most likely take advantage of technologies such as digital assistants that can automate the provision of knowledge through rule-based systems.

HiC-LoT PSFs will face limited pressures early on due to the tacit nature of the knowledge used in service provision. These firms are likely to incorporate emerging technologies to augment the services they offer as the richness and complexity of their operating environments increase (e.g. to process large sets of data reservoirs). These PSFs are more likely to increase their investments in the non-technical aspects of their service delivery approaches to limit the opportunity for their services to be viewed as a commodity.

LoC-HiT PSFs will face severe pressures to retain their competitive advantages. Specifically, these PSFs will have to find ways to use their existing knowledge bases to not only provide higher value services, which might require them to customize offerings, but additionally, to retain margins on standard services, they will need to stay ahead of competitors. Competitors might emerge from the IT sector in the form of start-ups who will aim to fully automate most, if not all, of the current set of standardized service offerings. As a result the margins on the provision of these services will shrink drastically thereby leading to significant organizational disruption.

HiC-HiT PSFs will face some disruption to their business models, but only if they (1) reduce their current level of investment in technology innovation, and (2) fail to retain their exclusive (leading) brand image due to their specialized tacit and relational knowledge. Disrupting these PSFs requires one to not only provide a radically new technology offering but also to scale up domain knowledge that is not explicated.

A PSF does not necessarily need to belong exclusively in one of the four cells. Larger PSFs might have practices (divisions/initiatives) in each of the four cells. For example they could derive 80% of their revenue from practices that provide largely cookie-cutter solutions where there is high-use of stored knowledge assets to provide technological solutions (LoC-HiT PSF). The rest of their revenue could be attributed to highly customized services for a select group of clients where the knowledge assets being employed are highly tacit and relational in nature (LoC-HiT PSFs).

A PSF could also have practices where the type of service to be delivered will straddle multiple quadrants (e.g., legal/financial advice for an initial public offering (IPO) on the stock market which requires a mix of complex tacit knowledge and streamlined repetitive information lodgment). Or there is also the option of using an established or impromptu network of PSF firms to provide the required combination of customization and technology. This could assist smaller PSFs to overcome the impracticality of investing in technologies for which they do not have the scale and service volume to realize an acceptable return on investment.

There is a huge diversity of PSFs with some firms offering a wide range of services while others specialize in a specific service for a particular industry which will influence how they can deal with new, data-driven services and business models. To reflect this diversity, we also categorize PSFs by the range of industries served, and degree of focus of services they offer, which relates directly to the value proposition and customer segments of the business model and as such will also influence the other components (matrix II in Fig. 1). MaS-MaI PSFs offer a broad portfolio of services (e.g. auditing, information technology consulting, legal, etc.) across multiple industry sectors (e.g. defense, manufacturing, etc.). FeS-FeI PSFs offer a limited range of services to a single or smaller set of industry sectors.

FeS-FeI PSFs will face significant pressures to stay ahead of digital disruption unless they meet the following criteria: (1) increase their investments, and therefore, the value of the tacit knowledge and relational capital they have with their clients in order for them to deeply personalize service experiences, and (2) leverage existing and emerging technologies to augment high-value added services and automate routine services. These firms might begin offering some of the services they previously charged a premium for, at highly discounted rates or even free as disincentive to likely new entrants disrupting them. (3) create industry unique data-driven digital partner networks with their key clients and/or strategic industry partners to increase the level of entry for competitors.

MaS-FeI PSFs are one-stop shops for services for specific industries. MaS-FeI PSFs service all functional areas from legal, to accounting, human resources, and IT. These firms will face significant pressures in the future to maintain their breadth of service coverage due to digital disruption. The most likely strategies for these firms to consider are to (1) offload low value services or outsource their provision to low-cost centers that can take advantage of automation, (2) increase the intensity of use of emerging technologies and data-driven business models to enhance value offerings in high margin services, (3) create industry unique data-driven digital partner networks with key clients to increase the level of entry for competitors, and (4) begin to envisage how they might exploit future trends that will reshape the industries they serve – i.e. look at the mega-data and technology trends impacting the industry and identify ways to align offerings with these trends and even lead the conversation on the future of the industry.

FeS-MaI PSFs offer a few specialized services to a large number of industries, examples include security management firms. These firms will stay ahead of digital disruption by increasing their engagement with the existing data they have about the range of industries they operate in to identify higher value services they can design. In addition, these firms are more likely to engage in acquisitions or alliances with start-ups who are building technologies relevant to their service offerings. Given the fact that these organizations can scale emerging technologies across industries due to existing relationships, the acquisition of, or alliance with, start-ups is a viable strategy (start-ups who have technologies need access to markets for their products).

MaS-MaI PSFs are multi-service firms who operate across a range of industries. These firms will take on digital disruption by pursuing the following strategies: (1) looking at quick ways of automating services that are common across a range of industries to enable economies of scale and quick recoupment of technology investments, (2) looking for new bundling of services that can add value through innovative data-driven offerings (e.g., look to connect previously one-off service offerings with a data-driven solution), and (3) liaising with existing start-ups to get access to new technologies and/or beginning to invest in their own R&D labs around new technologies.

5 Transitionary Pathways for PSFs

PSFs, like all organizations, will need to evolve based on changes in their environments (Meyer 1982; Siggelkow 2002). We now describe possible transitionary paths for PSFs to embrace digitization to transform their business models and operations following the strategic framework described below (Fig. 2).

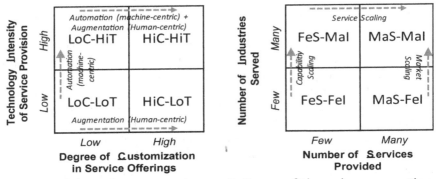

Fig. 2. Transitionary pathways for professional services.

We first discuss the pathways related to the level of customization and technology intensity of the service/s they provide (matrix I in Fig. 1). Given the advances brought on by digitization the number of LoC-LoT PSFs within a given industry will decrease. The future viability of setting up a LoC-LoT PSFs will be limited. Depending on their approach to change, a LoC-LoT PSFs can either choose an *incremental* or *radical* approach to change. An incremental approach would see a LoC-LoT PSFs take advantage of technological advances to augment its current workforce thereby increasing the level of customization of service offerings for its existing clients. This strategy would build on the existing social and relational capital (Bourdieu 1986) held by the firm and enhance it through delivery of personalized service offerings. The ability of a firm to build on its existing social and relational capital should increase the switching cost for existing clients when considering other lower cost alternatives. A radical approach will involve a LoC-LoT PSFs transforming its fundamental business model to one that is fully focused on data and takes advantage of technological offerings. In this case, the LoC-LoT PSFs will automate most, if not all of its current service offerings and re-structure (most likely downsize) its current workforce. The LoC-LoT PSFs will look to seek advantages of scale through its technological offering and then use its existing technological offerings to increase the breadth of its service offerings. It is important to note that the focus here is on using technology to automate services that are generic and can be completely handled by machines.

LoC-HiT PSFs have one likely scenario to upgrade their existing business models. They can simultaneously conduct automation for customizing and personalizing low-value services that are structured (or semi-structured) and then augment their experts through data-driven solutions that can sift through large data reservoirs, predictive models that can alert to new business/sales opportunities based on client developments, etc. These firms are likely to survive if they (1) leverage their current investments in their digital infrastructure, and (2) are able to swiftly re-orient the organization to move away from low-value added projects/jobs towards higher-value services. This requires an effort into both business model innovation and digital transformation.

HiC-LoT PSFs have one likely scenario to continue their relevance in the marketplace – leverage technologies to scale current customized service offerings, i.e., create customer-unique value through mass customization (Gilmore and Pine 2000). These firms will need to find ways to leverage technology towards providing personalized service offerings for their existing clients at scale. The focus here will be use the expertise found in their knowledge workers and design appropriate automated solutions. For example, with the rise of digital employees and robotics process automation (RPA), HiC-LoT PSFs have options to customize service deliveries using the former and scale service solutions that are generic using the latter. HiC-LoT PSFs will only survive if they can retain what Starbuck (1992) calls exceptional and valuable expertise, much of this will come down to deep client knowledge, social capital, and trust between the parties due to the sensitive nature of projects/transactions that the PSF will be undertaking.

HiC-HiT PSFs will continue to operate in the marketplace so long as they are able to leverage their current and new data-driven business models to design, make legitimate, and scale new service offerings. As a historical reference, consider the case of the 'Big 8' accounting firms that eventually morphed into the 'Big 4' PSFs of today. These organizations began as traditional accounting and auditing firms but moved into the assurance and information technology domain. Today, these are full-service management consulting firms that continue to invest in cannibalizing and legitimizing new service lines (Suddaby and Greenwood 2001) in order to retain their relevance and extend their brand and knowledge capabilities into new areas.

We next discuss the pathways related to the number of services and industries (matrix II in Fig. 2). FeS-FeI PSFs will need to take advantage of technologies to scale their current services into new industries. Put differently, they will have to find ways to (1) focus on their truly innovative market offerings, and (2) leverage technologies to enable them to enter, and advance their positions, in new industry segments. Consider the case of a boutique legal firm whose focus is on domestic merger and acquisitions litigation and assurance services operating in Cambridge, Massachusetts. The firm will have to find ways to take their existing M&A knowledge and enter international markets, or if they were focused on one industry (e.g., technology) find ways to port their existing knowledge into related spaces (e.g., defense, pharmaceuticals). Investments into technology will help them to develop digitized solutions for these new industry segments, while also adding new offerings for their existing clients. It is unlikely that FeS-FeI PSFs will have the necessary scale or slack resources to devise completely new services due to limited knowledge base and relational capital.

MaS-FeI PSFs should leverage their investments in technology to scale their service offerings across industries. These firms will be in a similar position as FeS-FeI

PSFs, with the key difference being is that they have a larger portfolio of services to take to new industries. It is highly likely that these firms will go through a process of service rationalization whereby existing services are prioritized and evaluated in terms of creating service bundles that can serve multiple industries as packaged offerings.

FeS-MaI PSFs will need to strategize their current service offerings from the potential of technological enhancements. Put differently, can a data-driven business model focus enable them to enhance (through add-ons) existing offerings to increase their value. In addition, given their broad knowledge of a wide assortment of industries, what kinds of new services might be created that have immediate access to a large enough client base to underwrite initial investments into technological and development costs.

The Dominant Pathways – Clearly one might argue that a given PSF can choose multiple pathways to advance from its current position. While, this might be true, in theory, we are focused on the dominant or most likely strategy. The simple reason for this is as follows: given the market pressures, intensity of competition, current investments in digital capabilities and infrastructure within the firm, and the state of slack resources to fund new initiatives, we assert that only HiC-HiT PSFs and MaS-MaI PSFs will have the luxury of experimenting with multiple options when it comes to upgrading their data-driven business models. These firms due to legacy investments in technology, human and client capital, and market positionings have the necessary resources to sustain continued investments in renewing their data-driven business models and leveraging to enter, or define, new markets.

6 Concluding Remarks

In this conceptual paper, we took a closer look at digital business model innovation in relation to PSFs. We first contrasted the traditional, human-centered, knowledge-intensive business model of PSFs with the new, computer-centered, data-driven business model that is developing due to the rise of big data, advanced data analytics and artificial intelligence. Next, to better understand if, when and how new data-driven business models may disrupt traditional knowledge-intensive business models, we provide an framework for identifying and analyzing the strategic options of PSFs addressing (1) the nature of their value proposition: the level of customization and technology intensity of the service/s they provide and (2) the scope of their value proposition: the degree of focus of services they offer and the range of industries served. Based on this strategic framework, we suggest a number of possible transitionary pathways using digital technology for augmentation or automation and the need so scale across services and industries.

PSFs are an interesting class of organizations to study in relation to digital innovation due to their knowledge-intensive nature. The survival of PSFs in the age of cognitive computing, robotic process automation, and real-time analytics comes down to a firm's ability to develop a viable computer-centered, data-driven business model that can complement and/or compete with traditional business models. Our insights can inform practitioners as they need to assess the impact of new digital technologies, business models and competitors on their organizations, customers, business networks and industries and develop new strategies and business models to deal with this.

The ideas presented in this paper need to be further developed and tested. Moreover, although the choices we made in relation to the classification schemes seem to result in valuable strategic insights, others may argue for the use of other classifications. For future work, we propose to conduct empirical studies to further explore the rise of data-driven business models within PSFs to better understand digital innovation contextualises to particular industries and the strategic options within those industries.

References

Baller, S., Dutta, S., Lanvin, B. (eds.): The Global Information Technology Report 2016: Innovating in the Digital Economy. World Economic Forum, Geneva (2016)

Bourdieu, P.: The forms of capital. In: Richardson, J.G. (ed.) Handbook of Theory and Research for the Sociology of Education, pp. 241–258. Greenwood Press, New York (1986)

Brynjolfsson, E., McAfee, A.: The Second Machine Age: Work, Progress, and Prosperity in a Time of Brilliant Technologies. W. W. Norton & Company, New York (2014)

Chen, H., Chiang, R.H.L., Storey, V.C.: Business intelligence and analytics: from big data to big impact. MIS Q. **36**, 1165–1188 (2012)

Chesbrough, H.: Open Business Models: How to Thrive in the New Innovation Landscape. Harvard Business School Press, Boston (2006)

Chesbrough, H.: Business model innovation: opportunities and barriers. Long Range Plan. **43**, 354–363 (2010)

Chin, J.K., Hagstroem, M., Libarikian, A., Rifai, K.: Advanced Analytics: Nine Insights from the C-suite. McKinsey & Company, New York (2017)

Cook, D.P., Goh, C.-H., Chung, C.H.: Service typologies: a state of the art survey. Prod. Oper. Manag. **8**, 318–338 (1999). https://doi.org/10.1111/j.1937-5956.1999.tb00311.x

DaSilva, C.M., Trkman, P., Desouza, K., Lindič, J.: Disruptive technologies: a business model perspective on cloud computing. Technol. Anal. Strateg. Manag. **25**, 1161–1173 (2013). https://doi.org/10.1080/09537325.2013.843661

Demil, B., Lecocq, X., Ricart, J.E., Zott, C.: Introduction to the SEJ special issue on business models: business models within the domain of strategic entrepreneurship. Strateg. Entrepreneurship J. **9**, 1–11 (2015). https://doi.org/10.1002/sej.1194

Empson, L., Muzio, D., Broschak, J., Hinings, B.: Researching professional service firms: an introduction and overview. In: Empson, L., Muzio, D., Broschak, J., Hinings, B. (eds.) The Oxford Handbook of Professional Service Firms, pp. 1–22. Oxford University Press, Oxford (2015)

Fichman, R.G., Dos Santos, B.L., Zheng, Z.E.: Digital innovation as a fundamental and powerful concept in the information systems curriculum. MIS Q. **38**, 329–353 (2014)

Fielt, E.: Conceptualising business models: definitions, frameworks and classifications. J. Bus. Models **1**, 85–105 (2013)

Fielt, E., Gregor, S.: What's new about digital innovation?. In: Paper presented at the Information Systems Foundations (ISF) Workshop, Canberra, AU (2016)

Foss, N.J., Saebi, T.: Fifteen years of research on business model innovation: how far have we come, and where should we go? J. Manag. **43**, 200–227 (2017)

Gilmore, J.H., Pine, B.J. (eds.): Markets of One: Creating Customer-Unique Value through Mass Customization. Harvard Business Review Book Series. Harvard Business School Publishing, Boston (2000)

Greenwood, R., Hinings, C.R.: Understanding strategic change: the contribution of archetypes. Acad. Manag. J. **36**, 1052–1081 (1993). https://doi.org/10.5465/256645

Greenwood, R., Suddaby, R., McDougald, M.: Introduction. In: Greenwood, R., Suddaby, R. (eds) Professional Service Firms (Research in the Sociology of Organizations, volume 24). Emerald Group Publishing Limited, pp. 1–16. (2006). https://doi.org/10.1016/s0733-558x(06)24001-1

Günther, W.A., Rezazade Mehrizi, M.H., Huysman, M., Feldberg, F.: Debating big data: a literature review on realizing value from big data. J. Strateg. Inf. Syst. **26**, 191–209 (2017). https://doi.org/10.1016/j.jsis.2017.07.003

Hartmann, P.M., Zaki, M., Feldmann, N., Neely, A.: Capturing value from big data: a taxonomy of data-driven business models used by start-up firms. Int. J. Oper. Prod. Manag. **36**, 1382–1406 (2016). https://doi.org/10.1108/IJOPM-02-2014-0098

Katz, D.M., Bommarito, M.J., Blackman, J.: A general approach for predicting the behavior of the supreme court of the United States. PLoS One, 12 (2017). http://dx.doi.org/10.1371/journal.pone.0174698

Loebbecke, C., Picot, A.: Reflections on societal and business model transformation arising from digitization and big data analytics: a research agenda. J. Strateg. Inf. Syst. **24**, 149–157 (2015)

Lovelock, C.H.: Classifying services to gain strategic marketing insights. J. Mark. **47**, 9–20 (1983)

Løwendahl, B.R., Revang, Ø., Fosstenløkken, S.M.: Knowledge and value creation in professional service firms: a framework for analysis. Hum. Relat. **54**, 911–931 (2001). https://doi.org/10.1177/0018726701547006

Maister, D.H.: Managing The Professional Service Firm Free Press, New York (1993)

Massa, L., Tucci, C.L.: Business Model Innovation. In: Dodgson, M., Gann, D., Phillips, N. (eds.) The Oxford Handbook of Innovation Management, pp. 420–441. Oxford University Press, Oxford (2014)

Massa, L., Tucci, C.L., Afuah, A.: A critical assessment of business model research. Acad. Manag. Annals **11**, 73–104 (2017). https://doi.org/10.5465/annals.2014.0072

Meyer, A.D.: Adapting to Environmental Jolts. Adm. Sci. Q. **27**, 515–537 (1982)

Morris, M., Schindehutte, M., Allen, J.: The entrepreneur's business model: toward a unified perspective. J. Bus. Res. **58**, 726–735 (2005)

Nambisan, S.: Information technology and product/service innovation: a brief assessment and some suggestions for future research. J. Assoc. Inf. Syst. **14**, 215–226 (2013)

Nambisan, S., Lyytinen, K., Majchrzak, A., Song, M.: Digital innovation management: reinventing innovation management research in a digital world. MIS Q. **41**, 223–238 (2017)

Newell, S., Marabelli, M.: Strategic opportunities (and challenges) of algorithmic decision-making: a call for action on the long-term societal effects of 'datification'. J. Strateg. Inf. Syst. **24**, 3–14 (2015). https://doi.org/10.1016/j.jsis.2015.02.001

Osterwalder, A., Pigneur, Y.: Business Model Generation: A Handbook for Visionaries, Game Changers, and Challengers. John Wiley and Sons, Hoboken (2010)

Schüritz, R., Seebacher, S., Satzger, G., Schwarz, L.: Datatization as the next frontier of servitization: understanding the challenges for transforming organizations. In: Proceedings of the 38th International Conference on Information Systems (ICIS 2017). Seoul, South Korea (2017)

Shah, S.: How blockchain is revolutionising the legal sector. Raconteur (2018)

Siggelkow, N.: Evolution toward fit. Adm. Sci. Q. **47**, 125–159 (2002)

Starbuck, W.H.: Learning by knowledge-intensive firms. J. Manag. Stud. **29**, 713–740 (1992)

Suddaby, R., Greenwood, R.: Colonizing knowledge: commodification as a dynamic of jurisdictional expansion in professional service firms. Hum. Relat. **54**, 933–953 (2001)

Westerman, G., Bonnet, D., McAfee, A.: Leading Digital: Turning Technology into Business Transformation. Harvard Business Review Press, Boston (2014)

Wixom, B.H., Ross, J.W.: How to monetize your data. Sloan Manag. Rev. **58**, 10–13 (2017)

Woerner, S.L., Wixom, B.H.: Big data: extending the business strategy toolbox. J. Inf. Technol. **30**, 60–62 (2015)

Yoo, Y., Henfridsson, O., Lyytinen, K.: The new organizing logic of digital innovation: an agenda for information systems research. Inf. Syst. Res. **21**, 724–735 (2010)

Impacts of Information Provision and Visualization on Collaborative Decision Making

Anh Luong[✉], Karl R. Lang, and Shadi Shuraida

Baruch College, City University of New York, New York, NY 10010, USA
{anh.luong,karl.lang,shadi.shuraida}@baruch.cuny.edu

Abstract. Big Data marks one of the biggest opportunities and also challenges of modern day organizations. Not all data is equal–only some contain insights. To make sound decisions, firms need not only process data, but also extract meaningful information, and present it in such a way that can illuminate perceptions. Intrigued by this topic's current challenges, this study examines how different manners of providing information affect collective decision-making performance. To examine the research question, an electronic platform in the laboratory is designed to implement a collaborative decision-making experiment using game theory. Experimental economics is used as the main research methodology, featuring financial incentives in the experiments in order to induce subjects' rationally strategic behaviors. The research wishes to yield meaningful insights that bear implications for many contexts, most important of which inter- and intra-organizational decision making, as well as group collaboration and coordination.

Keywords: Collaborative decision making · Information provision · Information visualization

1 Introduction

The present study examines the impacts of information provision and visualization schemes on collaborative decision-making performance. Specifically, it examines such impacts in the case of organizational decision-making on shared resource investments—investments made by multiple units, which, if successful, can result in a resource to be shared among the units. In both the general context of collaborative decision making and the specific context of organizational decision-making aforementioned, many factors can be said to affect decision quality, to name a few: the decision makers' competence, the dynamics among the deciding units, the nature of the decision problem, among many others. Flowing among and tightly integrated into these factors, is the information representing the decision problem that is provided to the decision makers. Only through information can the decision problem be communicated to the decision makers

© Springer Nature Switzerland AG 2019
J. J. Xu et al. (Eds.): WEB 2018, LNBIP 357, pp. 39–45, 2019.
https://doi.org/10.1007/978-3-030-22784-5_4

and can the decision makers' actions along with their wide ranging impacts be transmitted among themselves. It has been established that information plays a crucial role in organizational decision-making and information systems success (DeLone and McLean 1992). Hence, it is important that information is conveyed truthfully and effectively to decision makers (Ballou and Pazer 1985; Wang and Strong 1996; Zmud 1978).

Not all information is accurate or useful; even when such quality information is obtained, conveying it so as to minimize loss of meaning during the process is another challenge. Information Theory (Shannon 1948) states that the same information can be presented in different ways and different formats and human decision makers may process information differently depending how it is presented to them. Thus, individual and group decision quality may be affected by not just what type of information that is available but also by the way in which it is presented. In this day and age, with the explosion of big data and machine learning advances (Brynjolfsson and McAfee 2017), organizations need to be able to leverage their data visualization tools so that they can extract useful information from their data, and present it in a way that can elicit insightful interpretations from key decision makers and leaders. Otherwise, their big data resources will likely go to waste. Motivated by this interesting challenge, the present study investigates the impacts of various information visualization formats and provision schemes on collaborative decision making outcomes, with a focus on shared resource investment problems. This topic has been extensively studied, however, with the focus mostly on individual decision-making performance (Keller and Staelin 1987; Malhotra 1982; Park et al. 2016; Speier et al. 2003; Tam and Ho 2006; Todd and Benbasat 1999). Our study wishes to extend this literature by examining the topic in the context of collaborative decision making, which is vastly different from individual decision making, given that not only individuals' capabilities, but also group dynamics, exert considerable influence on the group decision outcomes.

2   Theoretical Foundation and Hypothesis Development

Having established that information plays a crucial role in the collaborative decision making, we consider the perspectives of Information Processing Theory (Miller 1956) and Cognitive Fit Theory (Vessey 1991) to explain decision makers' behaviors in a collaborative setting. In the context of recurring decisions making on shared resource investments, decision makers typically use the information that is available to them–for instance, what they, their group and each of the group members invested and earned during the previous period–to compose a mental model in order to discern patterns and trends to help them make (better) decisions for the next period. One would expect that the more information that is available to the decision makers in such a case, the easier it would be for them to analyze their partners' behavior trends and strategize their course of actions. On the other hand, too much information might overload cognitive processing and hinder decision making, as proposed by Information Processing Theory

(Miller 1956). Additionally, Miller (1956) claims that, since we humans can typically only store approximately only five to nine units of information in their working memory at a time, we are inclined to process information through the use of "chunking"–grouping large amounts of information into chunks–each of which contains bits of information similar or related to each other. Thus, **recoding**–aggregating information into fewer and larger chunks, is proposed to increase information processing efficiency. Hence, we posit in our first hypothesis that in a group decision making context, aggregate information will lead to better group decision outcomes, compared to fully detailed information. Additionally, as the number of decision makers in a group increases, the amount of information to be reported to and processed by the decision makers also increases. Thus, we also propose that the bigger the group, the stronger the positive effect of aggregating information on decision outcomes.

Hypothesis 1a. *Providing decision makers with aggregate information will lead to better* **group** *decision outcomes, compared to providing them with non-aggregate information.*

Hypothesis 1b. *For larger groups, the impact of providing decision makers with aggregate information on* **group** *decision outcomes is stronger.*

Further, Cognitive Fit Theory (Vessey 1991) posits that users' problem-solving performance improves when the physical presentation of a problem matches the nature of that problem and the skills it requires. Applying the theory to comparing decision makers' performance on *spatial* tasks (extracting discrete and precise data values) versus *symbolic* tasks (observing trends or relationships, and making associations), Vessey (1991) found that tables enabled decision makers to make faster and more accurate decisions on symbolic tasks than graphs did. Additionally, although graphs allowed users to complete spatial tasks faster than tables, the decisions made by the graph users were also less accurate. In the context of recurring decisions making on shared resource investments, the decision task involves both spatial and symbolic components. Since previous research on Cognitive Fit has found that different task representation formats (Table, Graphic, etc.) have differing impacts on individual decision outcomes, we posit that these differing impacts will also extend to group decision outcomes. Hence, we propose our second hypothesis as follows.

Hypothesis 2a. *Providing decision makers with information presented in different formats (Graphs, Tables, Raw Data) matters for* **group** *decision making outcomes.*

Similar to hypothesis 1, we also propose that as the group size increases, more information has to be processed by the decision makers, and thus, the differing impacts of various information formats on group decision outcomes will become stronger.

Hypothesis 2b. *For larger groups, the impact of providing information presented in different formats on* **group** *decision outcomes is stronger.*

3   Study Design

3.1   The Decision Problem

For the modeling of the collaborative decision making context that we are interested in, *shared resource investments*, we borrow from a reference discipline, Experimental Economics, the game theoretic insights of the public goods experiment. Public goods experiments have been used extensively in various fields to study issues in collaborative resource contribution and allocation. The term *public good* denotes any type of common resource (e.g. money, space, benefits, services) that is both *nonrival* and *nonexcludable*, meaning multiple people can consume the common resource simultaneously, and both those who do and those who do not contribute to the common resource are equally able to consume it, respectively. Public goods experiments typically involve decision makers being assigned into groups, each given some initial amount of resources (expressed in tokens or points) and required to allocate such resources between two accounts—their private and their group accounts. Each decision maker's earnings are the sum of their private accounts and their portion of returns from the group account, computed based on a predetermined rate.

Many variations of public goods experiments exist, each designed to study a specific context or a class of context (for a comprehensive review, see Davis and Holt 1993). For our study, we use the Provision Point Mechanism (PPM), as it matches our case of examination–shared, organization-wide funds that need a threshold (provision point) to be set up and distributed. Additionally, since we wish to examine decision makers' behaviors under a dynamic perspective, and given that organizational decision-making typically occurs in a recurring manner, we implement and analyze the multiple-round version of the public goods experiment (as opposed to the one-shot version, where there is only one decision period).

3.2   Experimental Design

In order to examine how different ways of providing and visualizing information impact collaborative decision outcomes, we design an electronic platform in the laboratory. We adopt the experimental economics methodology (e.g. Smith, 1976), which has increasingly been used in the Information Systems field (Gupta et al. 2018; Adomavicius et al. 2013; Bapna et al. 2010; Cason et al. 2011; Hashim et al. 2016; Rice 2012). The methodology is rigorous for examining humans' decision making process when facing complex strategic tasks. We adapt the parameters from a well-cited implementation by Bagnoli and McKee (1991). The study has a factorial design with two factors: **Information Visualization** (Raw Data vs. Tablular vs. Graphic), and **Information Provision** (Aggregate vs. Non-Aggregate). In all treatments, each participant is given feedback after each round about how much his/her group contributed and how much s/he earned in that round. In the Aggregate treatments, besides that information, participants are provided with their group average, minimum and maximum values of contribution, earnings, and time to make a decision. In the Non-Aggregate treatments, participants are provided with information of each of the group members' (including themselves) regarding contribution, earnings, and time to make a decision. In each session, participants are randomly assigned to groups of 10 to engage in the decision-making experiment for 25 rounds (the number of rounds undiclosed to participants to prevent last-round irregular behaviors). Before the experiment starts, the instructions are explained to them. Then, they start making decisions and viewing the results after each round until all of the rounds have passed. The performance-based earnings on average for participants are $15. The use of monetary incentive is meant to induce rationally strategic behavior, following Induced Value Theory (Smith 1976). Formally, the pay-off function for each of the 10 participants in our version of the game is the following:

$$p_i = \begin{cases} e_i - c_i, & g < 25 \\ e_i - c_i + r, & g \geq 25 \end{cases} \tag{1}$$

where p_i denotes each player's pay-off value, e_i the initial endowment, c_i each player's contribution in tokens, r the reward, and g the amount of the group account in each period. The experimental platform is implemented with the open source *oTree* software (Chen et al. 2016).

4   Data Analysis

We examine several measures of decision making performance: (1) groups' ability to (implicitly) coordinate around an equilibrium, measured by Distance to Equilibrium (DTE), and (2) groups' investment success rate (ISR). Importantly, we account for the round effects, separating and comparing the results of the early rounds with that of the later rounds. We are also aware of the inherent group dynamics and effects that might confound the experiment results, and thus, collect post-game questionnaire responses regarding the group cohesion degree within each group in order to potentially treat this as a control variable for the research model. We use regression analysis to test for the main effects, the possible interaction effects, and account for control variables, which include demographics information and personality trait constructs. These are asked in the pre- game and post-game questionnaires.

We have conducted experimental sessions using the diverse subject pool consisting of Business undergraduate students at Baruch College, City University of New York. The preliminary results are as follows.

First, the Group Distance to Equilibrium (DTE) differs among the six treatments noticeably. DTE is calculated as the absolute values of the difference between group accounts' values (g) and the equilibrium that is the closest to g (out of two equilibria, 0 and 25). Meaning, if the group account value is less than 12.5, the nearest equilibrium would be 0, and the DTE would be g. If otherwise, the nearest equilibrium would be 25, and the DTE would be $|g - 25|$. Since all of the groups that we analyzed yielded in all of the rounds group accounts values greater than 12.5, the DTE reported here are the deviations from the

25-point equilibrium, which is also the Pareto-efficient equilibrium. Treatments Raw Data-Aggregate (RA) & Table-Nonaggregate (TN) had the largest deviations; treatments Table-Nonaggregate (TN) and Graph-Nonaggregate (GN) had the smallest deviations; and treatments raw data-aggregate (RA) and graph-aggregate (GA) had medium deviations from the Pareto-optimal equilibrium.

Second, Investment Success Rate (ISR) is computed as the percentage of the rounds where group accounts reach at least 25 points. The only two treatments that did not have the 100% ISR were TN and GN. Thus, when combining these two group decision performance measures, DTE & ISR, we conclude that the two treatments that performed the best were RA and GA, as the groups in these treatments only had medium deviations from the Pareto-efficient equilibrium, and also yielded 100% investment success rate in all of the rounds. Although TN and GN had the smallest deviations from the Pareto-efficient equilibrium, they did not produce successful group investments in all of the rounds.

Third, regarding Regression Analysis: Factors found to have significant impact on the total earnings were information visualization and aggregation. Additionally, we found a small interaction effect that decreased earnings in the Raw Data condition, which could be attributed to the decision makers' pro-social trait. We also found an interaction effect that increased earnings in the Tabular condition, which could be attributed to the decision makers' computer self efficacy.

5   Conclusion

This study builds on an established body of literature that has looked at how individual decision makers use and process different information presentation formats and found that graphical information is often more effective than text-based information. Consistent with prior literature we find that data visualization and aggregation help groups coordinate better and make better decisions on collaborative tasks. The findings will bear implications for many contexts, most importantly for (inter- and intra-) organizational decision making, which increasingly occurs in online business settings where groups of managers look at shared information presentation in decision making process. Further, the study will contribute theoretically through expanding the current literature's understanding on individual information processing capabilities in collaborative settings with a dynamic perspective.

References

Adomavicius, G., Gupta, A., Sanyal, P.: Effect of information feedback on the dynamics of multisourcing multiattribute procurement auctions. J. Manage. Inf. Syst. **28**(4), 199–229 (2012)

Bagnoli, M., McKee, M.: Voluntary contribution games: efficient private provision of public goods. Econ. Inq. **29**(2), 351–366 (1991)

Ballou, D.P., Pazer, H.L.: Modeling data and process quality in multi-input, multi-output information systems. Manage. Sci. **31**(2), 150–162 (1985)

Bapna, R., Dellarocas, C., Rice, S.: Vertically differentiated simultaneous vickrey auctions: theory and experimental evidence. Manage. Sci. **56**(7), 1074–1092 (2010)

Brynjolfsson, E., McAfee, A.: The Business of Artificial Intelligence: What it Can-and Cannot-Do for Your Organization. Harvard Business Review, Brighton (2017)

Cason, T.N., Kannan, K., Siebert, R.: An experimental study of information revelation policies in sequential auctions. Manage. Sci. **57**(4), 667–689 (2011)

Chen, D.L., Schonger, M., Wickens, C.: oTree-an open-source platform for laboratory, online and field experiments. J. Behav. Exp. Finance **9**, 88–97 (2016)

Davis, D.D., Holt, C.A.: Experimental Economics. Princeton University Press, Princeton (1993)

DeLone, W.H., McLean, E.R.: Information systems success: the quest for the dependent variables. Inf. Syst. Res. **3**(1), 60–95 (1992)

Gupta, A., Kannan, K., Sanyal, P.: Economic experiments in information systems. MIS Q. **42**(2), 595–606 (2018)

Hashim, M., Kannan, K., Maximiano, S.: Information feedback, targeting, and coordination: an experimental study. Inf. Syst. Res. **28**(2), 203–449 (2017)

Keller, K.L., Staelin, R.: Effects of quality and quantity of information on decision effectiveness. J. Consum. Res. **14**(2), 200–213 (1987)

Malhotra, N.K.: Information load and consumer decision making. J. Consum. Res. **8**(4), 419–430 (1982)

Miller, G.A.: The magical number seven, plus or minus two: some limits on our capacity for processing information. Psychol. Rev. **63**(2), 81–97 (1956)

Park, H., Bellamy, M.A., Basole, R.C.: Visual analytics for supply network management: system design and evaluation. Decis. Support Syst. **91**, 89–102 (2016)

Rice, S.: Reputation and uncertainty in online markets: an experimental study. Inf. Syst. Res. **23**(2), 436–452 (2012)

Shannon, C.E.: A mathematical theory of communication. Bell Syst. Tech. J. **27**(3), 379–423 (1948)

Smith, V.: Experimental economics: induced value theory. Am. Econ. Rev. **66**(2), 274–279 (1976)

Speier, C., Vessey, I., Valacich, J.S.: The effects of interruptions, task complexity, and information presentation on computer-supported decision-making performance. Decis. Sci. **34**(4), 771–797 (2003)

Tam, K.Y., Ho, S.Y.: Understanding the impact of web personalization on user information processing and decision outcomes. MIS Q. **30**(4), 865–890 (2006)

Todd, P., Benbasat, I.: Evaluating the impact of DSS, cognitive effort, and incentives on strategy selection. Inf. Syst. Res. **10**(4), 356–374 (1999)

Vessey, I.: Cognitive fit: a theory-based analysis of the graphs versus tables literature. Decis. Sci. **22**(2), 219–240 (1991)

Wang, R.Y., Strong, D.M.: Beyond accuracy: what data quality means to data consumers. J. Manage. Inf. Syst. **12**(4), 5–34 (1996)

Zmud, R.W.: An empirical investigation of the dimensionality of the concept of information. Decis. Sci. **9**(2), 187–195 (1978)

Why Monetary Gift Giving? the Role of Immediacy and Social Interactivity

Bingjie Deng[(✉)] and Michael Chau

The University of Hong Kong, Pok Fu Lam, Hong Kong
amberdbj@connect.hku.hk, mchau@business.hku.hk

Abstract. Live streaming, as a new practice of online digital content business, generates revenue mainly through users' monetary gift giving to content providers. The designs of the live-streaming platforms allow users to experience real-time interactions with both content providers and the virtual community. This study investigates potential motivations regarding the immediacy and real-time social interactivity and tests their impacts on users' intention of monetary gift giving behavior. The findings will not only extend the self-presentation theory to new contexts but also contribute to the Internet gift economy by elucidating underlying motivations in the victual gifting cycle. The study will offer a new angle for online digital content businesses worldwide on how to construct interactive functions in online communities.

Keywords: Live streaming · Monetary gift giving · Immediacy ·
Social interactivity

1 Introduction

With technological advances, digital content businesses have started to seek innovative ways to generate revenue. A new business model called live streaming is becoming popular around the world. For example, on Twitch.tv, a live-streaming platform for video games launched in 2011, users can participate in chats with other users while watching the live streaming of online games. This platform has attracted 15 million daily active users [13] who contribute to its revenue via advertisements. Since 2015, the live-streaming business has been fast developing with several new functions. Typical ones include platforms or apps in China such as douyu.com, inke.cn, YY Live, hua-jiao.com, and huya.com. According to the Chinese research platforms, as of February 2018, the number of users of live-streaming platforms has increased to over 220 million, and the market penetration rate has been over 20%. These platforms create an online community for users to experience interaction with the content providers and other users, thus putting monetary gift giving into new contexts.

On live-streaming platforms, users can buy virtual currencies on the platforms through online payment, then exchange for virtual gifts at different price levels and "donate" the gifts to the content providers. When the streamer receives a virtual gift, he or she will get its value minus the 20–30% deducted by the platform and exchange the virtual currency back to RMB. Monetary gift giving is the mainstream of the streamers and platforms to get revenue. Different from traditional motivations of gift giving, users

© Springer Nature Switzerland AG 2019
J. J. Xu et al. (Eds.): WEB 2018, LNBIP 357, pp. 46–52, 2019.
https://doi.org/10.1007/978-3-030-22784-5_5

may have new desires to give virtual gifts on such platforms given the unique interaction designs. The first design is the interaction between users and streamers. Since the gifts are present on the screen, when the streamers receive gifts, they usually show thanks to the givers immediately, highlighting their names and the gifts, giving the giver acknowledgment or approval for his or her gifting behavior. It emphasizes on the giver's identity and changes his or her expectation for rewards. The other one is the interaction between users and the community. When there is a gift giving, the picture and the number of the gifts are shown on the screen to the community. Users will not only get recognition from the community by updating their profile but also sense the atmosphere of the community, such as whether peers in the community like to give gifts. Apart from these interactions, competition in the community may also relate to users' intentions of monetary gift giving. When users notice same-sex peers chasing after the target, they may change their desire to give [28].

Previous studies have attributed the attenuated social influence in virtual communities to the lack of immediacy [5]. However, living-streaming platforms characterized by real-time interactions, provide an opportunity to rethink the interpersonal behavior, such as the reciprocity process and users' social motives. Drawing on existing literature, we would like to ask (1) what factors will affect users' intentions of monetary gift giving in virtual communities and (2) how such factors interact with each other?

2 Literature Review and Hypotheses

2.1 Internet Gift Economy

The Internet has played a role in the gift economy, which based mainly on sharing [23, 27]. In the sharing process, reciprocity is required, and such interchange may be in the form of intangible rewards (e.g., enhanced reputation, desire for prestige) [9, 18, 22], or further leads to tangible rewards [32]. When the giver receives the receiver's feedback, the gift exchange is considered to be completed. Later, a new round of gift exchange will be created, thus enforcing the cycle of gifting. Prior literature has discussed the norm of reciprocity in gift giving behavior [12, 17, 29]. People may hold an "anticipated reciprocity" to get similar gifts back in exchange [22]. Reciprocity in gift giving is often delayed because it usually happens at a later stage. Moreover, relationships involved in gifting relationships are more related to moral economy logics [33].The personal dimension of such relationships is about identity, in which how the giver and recipient see each other matters. The other is the social aspect, which means to build and maintain social relationships. Traditional literature mainly focused on the individuals in the relationships [4], while recent studies have extended to a more complex system of relations regarding communities as a whole [10]. The giver may either receive an intangible reward from a gift–receiver or receive awards from the Internet community [32]. However, the immediacy in reciprocity and the rewards from the virtual community in the gifting process are still under-explored. In this study, the new features of instant feedback and status-updates on the live-streaming platforms will address such gaps.

2.2   Self-presentation Theory

Self-presentation is a behavior that aims to convey some information or image of oneself to other people [3]. Self-presentation theory is also explained as the reason to project a desired image of oneself to others, testing people's purchasing behavior in virtual communities [20]. The definitions indicate one crucial element of self-presentation, that is, the immediate peer group or the presence of the potential audience [15]. Groups may control rewards and punishments for their members because mutual interpersonal evaluations can shape human social interactions. As long as the group members continue to meet each other, they will be affected by such evaluations. Given the features and designs of live-streaming platforms, the monetary gift giving behavior can be regarded as a way of self-presentation under certain circumstance, since it includes self-image transformation and self-audience interactions. Moreover, such kind of monetary gift giving behaviour highlights a feature of real-time interaction, which is absent in the traditional contexts of self-presentation.

There are two types of self-presentational motivations: self-belief or self-disbelief [11]. A person may represent to others in a way he or she sincerely believes, or engage in a way lying to others. Later, researchers made it systematic by stating two sub-groups: acquisitive self-presentation and protective self-presentation [1]. The two motivational systems are separate and unrelated. The acquisitive style supports that people present themselves to get approval from their peers or groups. The protective style is in the service of avoiding disapproval, being associated with social anxiety, reticence, and conformity.

In acquisitive self-presentation, people are motivated to acquire benefits, such as respect, power, and reputation, from online interactivities [2]. Given the features of live-streaming platforms, we develop two variables based on the acquisitive self-presentational motivation. The interactive designs with streamers allow users to perceive their gift giving as an immediate rewarding behavior, such as recognition of their identities, or responses respecting and approving of the personalities they present. Therefore, we define *expectation of immediate reciprocity (EIR)* as a giver's expectation of rewards from the streamers right after the giving behavior. When individuals regard others' responses to be validating and understanding, they will make more considerable efforts to maintain the relationship or the social bond [30]. Furthermore, some studies use reciprocity as a representation of social interaction and test its effect on charitable giving decisions [19, 24]. In social media, individuals may well expect intangible benefits to affect their behaviour further [25]. Therefore, the expectation of immediate feedback from the gift receiver may stimulate the giver's intention to give. The other variable, *virtual status-seeking (VSS)*, the desire to increase personal status in the virtual community, is represented by real-time screen updates of users' profile. When they give more gifts, users' status can be upgraded on the screen, which represents the rewards from the community. The real-time status-upgrade displayed in the virtual community may motivate users to give. Studies have shown that effective social interaction motivates users' contribution and continuance intention in a virtual community [6, 34–36]. Therefore, we make hypotheses that in the context of real-time interaction, the two new variables (i.e., EIR, VSS), will motive users' intention of monetary gift giving behavior:

H1a: Expectation of immediate reciprocity has a positive impact on a user's intention of monetary gift giving.

H1b: Virtual status-seeking has a positive impact on a user's intention of monetary gift giving.

Protective self-presentation means people present themselves to be in alliance with the rules of the community, to prevent criticism and to get into the group. The image of self for self-presentation could be defensive, rather than claiming some aspects of self [15]. When the giving records are shown on the live-streaming screen through animations, users are able to sense the rules of the community. In this study, we use the term *overwhelming gifting signal (OGS)* to represent the pressure that comes from the high-intense gift giving behavior in the community, usually represented by exaggerated and eye-grabbing animations. People experience constraints from the expectations of fellow members might change their identities. Moreover, people may be driven by the desire to conform to the social norms to perform voluntary behaviour [26]. Moreover, strong normative pressure might evoke the mood of "have to" rather than "want to," which affects their contribution desires [8]. We propose that the desire to conform to peers' behavior highlighted by eye-grabbing animations, would promote users' intention to present themselves in the virtual community.

H2: Overwhelming gifting signal has a positive impact on a user's intention of monetary gift giving.

2.3   Competition

Competition very often exists in social media contexts [31]. Since more than 80% of the contributors to a content provider is from the opposite sex, live streaming is a typical context for the competition between same-sex users. When males are donating to an attractive female fundraiser, they respond competitively to donations made by other men. The researchers suggest a role for sexual selection in explaining remarkable generosity [28]. Based on social value orientation theory, an individual's heterogeneous preferences may result in competitive individuals [16]. Since users can detect a giver's gender on the interactive screen, we name the observation of gift giving behavior from same-sex givers as a user's perceived competition (PC). As discussed before, users may give gifts to the streamers to get immediate feedback or rewards from the community. When there is perceived competition in this process, the desire to acquire benefits may be stronger.

H3a: Perceived competition moderates the effect of expectation of immediate reciprocity on a user's intention of monetary gift giving.

H3b: Perceived competition moderates the effect of virtual status-seeking on a user's intention of monetary gift giving.

3   Research Model

This study uses expectation of immediate reciprocity (EIR), virtual status-seeking (VSS), and overwhelming gifting signal (OGS) as independent variables. Perceived competition (PC) acts as a moderator in the model. Meanwhile, there will be several control variables. Factors such as gender, age, education level, disposable income, online experience and online community participation activities affect giving behaviour and digital content consumption on social network sites [7, 14]. Types of content and user's perception of content will also be controlled (Fig. 1).

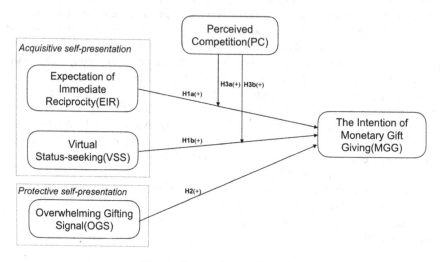

Fig. 1. Proposed research model

4   Methodology

We will conduct three lab experiments to test our hypotheses respectively. Specifically, the first experiment is a 2 (EIR vs. not, i.e., whether the giver receives the streamer's feedback immediately) by 2 (perceived competition vs. not, i.e., whether the same-sex competition occurs) between- subject design. The second experiment is a 2 (VSS vs. not, i.e., whether the status-upgrade is immediately displayed to the community) by 2 (perceived competition vs. not) between-subject design. The third one aims to test the effect of OGS (i.e., the animations along with gift giving) on users' intention of monetary gift giving. A field study will also be included to extend external validity of the results from lab study.

5   Ongoing Work and Potential Contributions

In this study, we investigate the role of interactive social factors in a user's intentions of monetary gift giving in the context of live streaming. With the interactive designs, live-streaming platforms in China provide us an opportunity to study personal and social dimensions of the monetary gift giving process. We test three variables, expectation of immediate reciprocity (EIR), virtual status-seeking (VSS) and overwhelming gifting signal (OGS) based on two subgroups of self-presentation theory. We also test how perceived competition affects users' giving decisions. The findings will have three main theoretical implications. First, it will primarily contribute to the self-presentation literature by exploring motivations regarding the immediacy and social interactive designs in virtual communities. Second, it will contribute to the gift economy literature by figuring out gifting motivations in an immediate and transparent cycle. Third, it will contribute to the literature related to competition and giving behavior, by providing an insight into the competition involved in real-time online interaction. This study also has several practical implications. For businesses utilizing monetary gift giving strategies or monetarizing digital content, managers and designers can adopt the designs of interactions, such as direct and immediate interaction with the content provider and information about real-time actions of other users in the community. For the content providers in a generalized scope, they may highlight the interactive atmosphere, the sense of being in the community, and the observed competition in the community to further speed up the giving cycles.

References

1. Arkin, R.M.: Self-presentation styles. In: Impression Management Theory and Social Psychological Research, pp. 311–334 (1981)
2. Barclay, P., Willer, R.: Partner choice creates competitive altruism in humans. Proc. R. Soc. Lond. B Biol. Sci. **274**(1610), 749–753 (2007)
3. Baumeister, R.F., Hutton, D.G.: Self-presentation theory: self-construction and audience pleasing. In: Mullen, B., Goethals, G.R. (eds.) Theories of Group Behavior, pp. 71–87. Springer, New York (1987). https://doi.org/10.1007/978-1-4612-4634-3_4
4. Belk, R.: Sharing. J. Consumer Res **36**(5), 715–734 (2009)
5. Chidambaram, L., Tung, L.L.: Is out of sight, out of mind? An empirical study of social loafing in technology-supported groups. Inf. Syst. Res. **16**(2), 149–168 (2005)
6. Chiu, C.M., Hsu, M.H., Wang, E.T.G.: Understanding knowledge sharing in virtual communities: an integration of social capital and social cognitive theories. Decis. Support Syst. **42**(3), 1872–1888 (2006)
7. Dou, W.Y.: Will internet users pay for online content? J. Advert. Res. **44**(4), 349–359 (2004)
8. Fu, F.Q., Richards, K.A., Hughes, D.E., Jones, E.: Motivating salespeople to sell new products: the relative influence of attitudes, subjective norms, and self-efficacy. J. Mark. **74**(6), 61–76 (2010)
9. Ghosh, R.A.: Cooking pot markets: an economic model for the trade in free goods and services on the internet. First Monday **3**(2) (2005) (originally published in March 1998)
10. Giesler, M.: Consumer gift systems. J. Consumer Res **33**(2), 283–290 (2006)
11. Goffman, E.: The presentation of self in everyday life. Garden City, NY (1959)
12. Gouldner, A.W.: The norm of reciprocity: a preliminary statement. Am. Sociol. Rev. **25**, 161–178 (1960)
13. Hamilton, W.A., Garretson, O., Kerne, A.: Streaming on twitch: fostering participatory communities of play within live mixed media. In: Proceedings of the SIGCHI Conference on Human Factors in Computing Systems (2014)
14. Han, B., Windsor, J.: User's willingness to pay on social network sites. J. Comput. Inform. Syst. **51**(4), 31–40 (2011)
15. Hogan, R.: A socioanalytic theory of personality. Paper presented at the Nebraska Symposium on Motivation (1982)

16. Huang, N., et al.: Motivating user-generated content with performance feedback: evidence from randomized field experiments. Manage. Sci. **65**(1), 327–345 (2018)
17. Joy, A.: Gift giving in Hong Kong and the continuum of social ties. J. Consumer Res. **28**(2), 239–256 (2001)
18. Kelty, C.: Free science/free software. First Monday **6**(12) (2001)
19. Khadjavi, M.: Indirect reciprocity and charitable giving-evidence from a field experiment. Manage. Sci. **63**(11), 3708–3717 (2017)
20. Kim, H.-W., Chan, H.C., Kankanhalli, A.: What motivates people to purchase digital items on virtual community websites? the desire for online self-presentation. Inf. Syst. Res. **23**(4), 1232–1245 (2012)
21. Klein, J.G., Lowrey, T.M., Otnes, C.C.: Identity-based motivations and anticipated reckoning: contributions to gift-giving theory from an identity-stripping context. J. Consumer Psychol. **25**(3), 431–448 (2015)
22. Kollock, P., Smith, M.: Communities in cyberspace. In: Communities in Cyberspace, pp. 13–34. Routledge, New York (2002)
23. Kube, S., Maréchal, M.A., Puppe, C.: The currency of reciprocity: gift exchange in the workplace. Am. Econ. Rev. **102**(4), 1644–1662 (2012)
24. Liu, C.J., Hao, F.: Reciprocity belief and gratitude as moderators of the association between social status and charitable giving. Personality Individ. Differ. **111**, 46–50 (2017)
25. Liu, Z.L., Min, Q.F., Zhai, Q.G., Smyth, R.: Self-disclosure in Chinese micro-blogging: a social exchange theory perspective. Inf. Manag. **53**(1), 53–63 (2016)
26. Martin, R., Randal, J.: How is donation behaviour affected by the donations of others? J. Econ. Behav. Organ. **67**(1), 228–238 (2008)
27. Offer, A.: Between the gift and the market: the economy of regard. Econ. Hist. Rev. **50**(3), 450–476 (1997)
28. Raihani, N.J., Smith, S.: Competitive helping in online giving. Curr. Biol. **25**(9), 1183–1186 (2015)
29. Sahlins, M.: Stone Age Economics. Routledge, New York (2017)
30. Schimel, J., Arndt, J., Pyszczynski, T., Greenberg, J.: Being accepted for who we are: evidence that social validation of the intrinsic self reduces general defensiveness. J. Pers. Soc. Psychol. **80**(1), 35–52 (2001)
31. Tang, Q., Gu, B., Whinston, A.B.: Content contribution for revenue sharing and reputation in social media: a dynamic structural model. J. Manage. Inform. Syst. **29**(2), 41–75 (2012)
32. Veale, K.: Internet gift economies: voluntary payment schemes as tangible reciprocity. First Monday **8**(12) (2003)
33. Weinberger, M.F., Wallendorf, M.: Intracommunity gifting at the intersection of contemporary moral and market economies. J. Consumer Res. **39**(1), 74–92 (2011)
34. Yoo, W.-S., Lee, Y., Park, J.: The role of interactivity in e-tailing: creating value and increasing satisfaction. J. Retail. Consum. Serv. **17**(2), 89–96 (2010)
35. Zhang, C.B., Li, Y.N., Wu, B., Li, D.J.: How WeChat can retain users: roles of network externalities, social interaction ties, and perceived values in building continuance intention. Comput. Hum. Behav. **69**, 284–293 (2017)
36. Zhao, L., Lu, Y.B.: Enhancing perceived interactivity through network externalities: an empirical study on micro-blogging service satisfaction and continuance intention. Decis. Support Syst. **53**(4), 825–834 (2012)

Revealing the Black Box of Privacy Concern: Understanding How Self-disclosure Affects Privacy Concern in the Context of On-Demand Services Through Two Competing Models

Chenwei Li[(⊠)] and Patrick Y. K. Chau

Faculty of Business and Economics, The University of Hong Kong,
Pok Fu Lam, Hong Kong
chenweli@connect.hku.hk

Abstract. As a prevalent economic paradigm, on-demand services match service providers and consumers with respective needs through the on-demand service platform. Consumers have to express their needs through self-disclosure, which inevitably raises privacy concern. However, how consumers' self-disclosure influences their privacy concern has not been well studied and remains as a black box. In this study, we would like to investigate how consumers' prior self-disclosure affects their privacy concern through two competing models derived from two theories in the literature: prominence interpretation theory and information processing theory. Based on prominence interpretation theory, the first model explains how the amount of consumers' prior self-disclosure in the past use affects the prominence and interpretation of requests for self-disclosure, thus finally influences consumers' privacy concern about their information. Based on information processing theory, the second model proposes a two-step approach that the amount of consumers' prior self-disclosure in the past use affects consumers' beliefs in the first step, and in the second step consumers' beliefs impact their evaluation of the on-demand service platform, thus finally influence their privacy concern. The models will be tested based on survey data collected from on-demand service consumers. The potential theoretical contributions and practical implications for consumers, service providers, and platforms are discussed.

Keywords: Privacy concern · Self-disclosure ·
Prominence interpretation theory · Information processing theory ·
On-demand services

1 Introduction

With the popularization of the Internet and the development of information technologies, people's names, ages, photos and consumption habits are becoming huge amounts of data stored in various forms. This trend of data into an important information resource makes consumers' privacy face a great risk. In-depth research on privacy concern needs to be conducted to help consumers relieve their privacy concern and share personal data. Scholars have long been interested in explaining the impact of

© Springer Nature Switzerland AG 2019
J. J. Xu et al. (Eds.): WEB 2018, LNBIP 357, pp. 53–62, 2019.
https://doi.org/10.1007/978-3-030-22784-5_6

privacy concern on consumers' behaviors from a rational perspective with privacy calculus theory as the main theoretical basis [1–3], positing that individuals are more likely to act if they consider the benefits are high enough to outweigh the costs [4, 5]. Among various antecedences of privacy concern, the request for self-disclosure may have a place. However, how consumers' self-disclosure influences their privacy concern has not been well studied and remains as a black box. Thus, we want to push this issue one step further in this study.

On-demand services have sprung up to match service providers and consumers with respective needs through the on-demand service platform. Examples include Airbnb, Uber, Amazon Mechanic Turk, Lyft, Homeaway, Mobike, etc. Meanwhile, driven by new information technologies, self-disclosure is closely associated with providing accurate and efficient services for consumers. Inevitably, it has also raised issues about the privacy and security of personal data. Much research has examined consumers' privacy concern in the context of social networking [6, 7], online transactions [8], and mobile applications [9, 10]. However, little attention has been paid to the privacy concerns in the on-demand services.

This study aims to investigate how consumers' self-disclosure influences their privacy concern in the context of on-demand services. We propose two competing models derived from two theories in the literature: prominence interpretation theory and information processing theory. Based on prominence interpretation theory, the first model explains the amount of consumers' prior self-disclosure in the past use affects the prominence and interpretation of requests for self-disclosure, thus finally influences their privacy concern. Based on information processing theory, the second model proposes a two-step approach that the amount of consumers' prior self-disclosure in the past use affects consumers' beliefs in the first step, and in the second step consumers' beliefs impact their evaluation of the on-demand service platform, thus finally influence their privacy concern.

The rest of the paper is structured as follows. Section 2 reviews previous literature on privacy concern and self-disclosure. Section 3 presents the theoretical foundation of the study, followed by Sect. 4, which presents the research models and corresponding hypotheses. Section 5 details the proposed methodology. Section 6 concludes with a discussion of potential contributions, implications and possible future directions.

2 Literature Review

2.1 Privacy Concern

Privacy concerns are worries about opportunistic use of personal information disclosed, which represent the degree to which individuals consider a privacy loss through the disclosure of personal information [11].

Prior studies on privacy concern mainly concentrate on the exploration of its influencing factors. Smith et al. [12] provided an interdisciplinary review of privacy research and concluded that consumers' personality differences will impact consumers' privacy concern. Other demographic differences, such as gender, age and education, have also been examined [8, 13–15]. Dinev and Hart [1] offered an understanding of privacy concern through the balance between privacy risk beliefs and privacy benefit beliefs, followed by other scholars [4, 16–18], such as personalization versus privacy intrusiveness, trust versus risks, media richness versus anonymity. Xu et al. [19] and Mousavizadeh et al. [20] studied the effects of privacy assurance approaches on information privacy concern. Platform reputation and platform familiarity were also documented [13].

Privacy calculus theory has dominated the analysis of privacy concern. For instance, Min and Kim [21] adopted the calculus of a cost–benefit framework and suggested privacy concerns as cost factors. When privacy risks were too pressing to offset potential benefits, users would limit their self-disclosure [22]. Privacy calculus theory could admittedly help understand user desire and user behavior if we treat them as rational consumers [4, 5]. However, not all users are rational, and a user cannot be rational all the time [22]. Seldomly, there was research examining consumers' privacy concern from a cognitive perspective of how it is perceived and formed. In this study, we propose two competing models derived from two theories in the literature, prominence interpretation theory and information processing theory, to study consumers' privacy concern in the context of on-demand services.

2.2 Self-disclosure

Self-disclosure refers to the act of voluntarily and intentionally disclosing any kind of information, such as addresses, hobbies and photos, to others when registering or using websites or mobile applications [3]. Prior research has demonstrated that users' self-disclosure may significantly influence users' self-disclosure through affecting their privacy concern. For instance, with the increase of the amount of information requested, individuals' privacy trade-off went negatively and intention to disclose their information decreased rapidly [23]. Anderson et al. [24] adopted privacy boundary theory to explain that the types of information requested and requesting stakeholders would alter individuals' privacy concern about disclosing themselves in healthcare settings. Individuals tended to have greater privacy concerns when sensitive information (e.g. financial, medical, demographic information) was requested [25].

Although researchers have spent effort in examining self-disclosure, the mechanism about how consumers' self-disclosure influences their privacy concern still remains as a black box. In this study, we would like to investigate how consumers' prior self-disclosure influences their privacy concern in the context of on-demand services. In particular, we would like to use two competing theories, namely prominence interpretation theory and information processing theory, to investigate how the amount of consumers' prior self-disclosure in the past use affects their privacy concern about using their information.

3 Theoretical Foundation

3.1 Prominence Interpretation Theory (PIT)

Prominence interpretation theory was developed as a way to understand how people make credibility assessments of websites [26]. The basic idea is that the success of website depends on whether users perceive the website to be credible. Given an external cue, two things happen when people assess the credibility of a website: they need to notice something (*prominence*); meanwhile, they make a judgment about what they notice (*interpretation*) [27].

Despite the various dimensions to access credibility [28–30], Fogg et al. [31] summarized two main components: *expertise* and *trustworthiness* when developing prominence interpretation theory. Following paper [31], credibility has been defined as "the extent to which the source is perceived as possessing expertise relevant and can be trusted to give opinion on the subject" [32], which indicates that expertise and trustworthiness are considered as antecedents of credibility. Therefore, we plan to adopt perceived expertise of requests for self-disclosure as a measure for the prominence of the requests, and consumers' perceived trustworthiness of on-demand service platform as a measure for the interpretation of the requests.

3.2   Information Processing Theory (IPT)

Information processing theory [33] was formulated to explain consumers' information processing of service delivered cues during the interactions between service consumers and service providers. Information processing theory suggests that consumers' information processing could be decomposed into four stages: *involvement experienced by consumers*, which varies with the amount and equivocality of information that the service platform requires from customers for service production, and the planned social interactions between service consumers and service providers; *consumers' expectation of involvement*, where consumers rely on the attributes of the service delivery process to generate expectation of the service; *confirmation or disconfirmation of consumers' beliefs*, where involvement experienced by customers is compared against consumers' expectation of involvement; and finally, an *evaluation* stage, where shows the consequences of information processing of the service delivered cues and consumers' evaluation towards the service.

In the on-demand services, involvement experienced by customers can be reflected by the amount of their self-disclosure in the past use. Consumers' perceived expertise is adopted to capture the consistency between requested self-disclosure and their past experience. Perceived trustworthiness is adopted to reflect consumers' evaluation of the service platform when the requests for self-disclosure are delivered during the service process.

4   Research Models and Hypotheses

4.1   Research Model Based on PIT

In Fig. 1, we present the research model overarched by prominence interpretation theory. It proposes that the amount of consumers' prior self-disclosure in the past use affects influences consumers' privacy concern through two routes: perceived expertise of requests for self-disclosure and perceived trustworthiness of on-demand service platform.

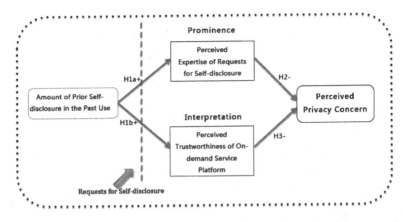

Fig. 1. Research model based on prominence interpretation theory

Effects of Amount of Prior Self-disclosure on Perceived Expertise Prominence is related with the experience of consumers, i.e., whether they are novice or expert [27]. For example, the aspects of a system that get noticed by an experienced individual will differ from what gets noticed by a novice user [34]. Consumers who have experience in the past use could better understand the purpose of the requests, and there is less possibility that requests for self-disclosure are perceived as deviated with their expectation. Therefore, those requests are less likely to be reviewed as obtrusive and task-irrelevant. Thus, we hypothesize that:

Hypothesis 1a: Amount of consumers' prior self-disclosure in the past use is positively related with consumers' perceived expertise of requests for self-disclosure.

Effects of Amount of Prior Self-disclosure on Perceived Trustworthiness Every individual has a certain basic level of skepticism toward a new request for personal information when insufficient information is involved [35], and they generally tend to remain skeptical thus degrade their initial trust level unless new information is entered [36]. Sufficient information provision for decision-making could greatly reduce consumers' level of skepticism and enhance trustworthiness [37]. When individuals have more experience of the service concerned, they are more likely to form positive beliefs towards the service, which seems to be more reliable and dependable to them [38]. Thus, we hypothesize that:

Hypothesis 1b: Amount of consumers' prior self-disclosure in the past use is positively related with consumers' perceived trustworthiness of on-demand service platform.

Effects of Perceived Expertise on Perceived Privacy Concern Expertise was proved to successfully drive social confidence [39]. When expertise is higher, competence perceptions tend to be stronger, individuals tend to have more confidence of control in their information [40], and worry less about their expectation would be violated. If a self-disclosure request is perceived as expertised, consumers will feel confident enough to trust its good intention thus won't concern about their privacy. Meanwhile, expertise could help enhance credibility [41]. A specialized technology would be perceived as more credible than a general one, and users feel more relieved to use the specialized one [42]. Likewise, an expertised request for self-disclosure is more likely to be credible and be accepted. Thus, we hypothesize that:

Hypothesis 2: Perceived expertise of request for self-disclosure is negatively related with consumers' perceived privacy concern about their information.

Effects of Perceived Trustworthiness on Perceived Privacy Concern Trustworthiness has been reported in the literature to hold a negative relationship with privacy concern [43]. When the source of information is highly trustworthy, individuals are more likely to attribute the piece of information to be beneficial and favorable [44]. Reasonably, individuals would develop greater affective regard and worry less about their perceived concern than those low in trustworthiness [45]. If consumers perceive the on-demand service platform as trustworthy, they will form beliefs that the platform is competent to manage their information, act properly and avoid opportunism use of their information. Thus, we hypothesize that:

Hypothesis 3: Perceived trustworthiness of on-demand service platform is negatively related with consumers' perceived privacy concern about their information.

4.2   Research Model Based on IPT

In Fig. 2, we present the research model overarched by information processing theory. It depicts that amount of consumers' prior self-disclosure in the past use is associated with perceived expertise of requests for self-disclosure that consumers believe, and then it affects consumers' perceived trustworthiness of the on-demand service platform, thus finally influences their privacy concern.

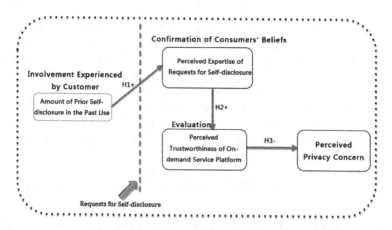

Fig. 2. Research model based on information processing theory

Effects of Amount of Prior Self-disclosure on Perceived Expertise Involvement experienced by customers is mainly embodied at the amount of consumers' prior self-disclosure in the past use [27]. The more involvement consumers experience in their encounters with service platform, the more motivated they are to attend to and comprehend the service delivered cues, and increase their tendency to perceive the goodwill of requested self-disclosure [33]. Meanwhile, under higher motivation, consumers are more competent and confident to accept the rationality of the requests [46], and perceived the requests as reasonable and professional. In other words, the likelihood for consumers to perceive a self-disclosure request as expertised will rise as their involvement level increases with their past self-disclosure. Thus, we hypothesize that:

Hypothesis 1: Amount of consumers' prior self-disclosure in the past use is positively related with consumers' perceived expertise of requests for self-disclosure.

Effects of Perceived Expertise on Perceived Trustworthiness Perceived expertise reflects consumers' confirmation of self-disclosure requests. When messages are expertised, receivers will consider them to be more constructive and persuasive [45]. Vincent and Webster [45] suggest that consumers' trust increases if they perceive a high level of expertise, because they will have more confidence of control and worry less about potential risks [39]. Meanwhile, beliefs of confirmation with requested self-disclosure help consumers form an expression that they can cooperate with the service platform [33]. When the requests are perceived as expertised, consumers' trust towards the platform will increase and they tend to give more credit to it. Thus, the perceived trustworthiness of on-demand service platform is enhanced. Thus, we hypothesize that:

Hypothesis 2: Perceived expertise of request for self-disclosure is positively related with consumers' perceived trustworthiness of on-demand service platform.

Effects of Perceived Trustworthiness on Perceived Privacy Concern Trustworthiness has been reported in the literature to hold a negative relationship with privacy concern [43]. Privacy concern can be viewed as consumers' worry about the chance that the service platforms use their confidential information without their permission [47]. When trustworthiness is higher, people are more likely to believe that a certain behavior is beneficial, and to evaluate the relevant parties more favorably [45], which may show less possibility to abuse their information. If an on-demand service platform is trustworthy, consumers will deem the platform as reliable, which won't abuse their confidential information without permission. Thus, we hypothesize that:

Hypothesis 3: Perceived trustworthiness of on-demand service platform is negatively related with consumers' perceived privacy concern.

5   Research Methodology

We target on-demand service consumers as our respondents. The selection criteria include previous experiences with on-demand services and current active use at the time of conducting survey. In the survey, respondents are required to indicate which pieces of information they have disclosed in the past use. We have listed several pieces of information that consumers may run into in their past use based on the general design of on-demand services, such as phone numbers, e-mail address, photos, real time locations, etc. After that, they need to report their perceptions about expertise, trustworthiness and privacy concern through seven-point Likert scales. Finally, demographic statistics including age, gender, education level, frequency of use and registration date will also be collected [48].

The amount of consumers' prior self-disclosure is measured by the total amount of information consumers have disclosed in the past when they were using the on-demand service app. Other items used to measure perceived expertise, perceived trustworthiness and perceived privacy concern are adapted from prior studies [8, 31]. They are based on seven-point Likert scales anchored from "strongly disagree" to "strongly agree" where 1 = strongly disagree and 7 = strongly agree.

6 Potential Contributions and Implications

This research is expected to contribute to the existing literature in several aspects. First, different with previous literature relying on privacy calculus theory to illustrate privacy concern, we offer a cognitive approach to assist in understanding how consumers' self-disclosure influences their privacy concern. Second, we extend previous literature about self-disclosure by considering how the requests for self-disclosure are received and interpreted by consumers as external service delivered cues, and the effect of prior self-disclosure on consumers' privacy concern. Third, we add to the body of prominence interpretation theory literature by contextualizing requests for self-disclosure as external cue and adapting it to investigate the relationship between privacy concern and its antecedents. We also add to the body of information processing theory literature by contextualizing requests for self-disclosure as service delivered cue and adapting it to investigate how the requests are processed to influence privacy concern.

References

1. Dinev, T., Hart, P.: An extended privacy calculus transactions model fore-commerce transactions. Inf. Syst. Res. **17**, 61–80 (2006)
2. Keith, M.J., Thompson, S.C., Hale, J., Lowry, P.B., Greer, C.: Information disclosure on mobile devices: re-examining privacy calculus with actual user behavior. Int. J. Hum Comput Stud. **71**, 1163–1173 (2013)
3. Posey, C., Lowry, P.B., Roberts, T.L., Ellis, T.S.: Proposing the online community self-disclosure model: the case of working professionals in France and the U.K. who use online communities. Eur. J. Inf. Syst. **19**, 181–195 (2010)
4. Jiang, Z., Heng, C.S., Choi, B.C.F.: Privacy concerns and privacy-protective behavior in synchronous online social interactions. Inf. Syst. Res. **24**, 579–595 (2013)
5. Koohikamali, M., Peak, D.A., Prybutok, V.R.: Beyond self-disclosure: Disclosure of information about others in social network sites. Comput. Human Behav. **69**, 29–42 (2017)
6. Li, K., Lin, Z., Wang, X.: An empirical analysis of users' privacy disclosure behaviors on social network sites. Inf. Manag. **52**, 882–891 (2015)
7. Liu, Z., Min, Q., Zhai, Q., Smyth, R.: Self-disclosure in Chinese micro-blogging: a social exchange theory perspective. Inf. Manag. **53**, 53–63 (2016)
8. Bansal, G., Zahedi, F.M., Gefen, D.: Do context and personality matter? trust and privacy concerns in disclosing private information online. Inf. Manag. **53**, 1–21 (2016)
9. Keith, M.J.,, Babbr, J.S. Jr., Furner, C.P., Abdullat, A.: Privacy assurance and network effects in the adoption of location-based services: an iphone experiment. In: ICIS 2010 Proceedings, p. 237 (2010)

10. Xu, H., Teo, H.-H., Tan, B.C.Y., Agarwal, R.: The role of push-pull technology in privacy calculus: the case of location-based services. J. Manag. Inf. Syst. **26**, 135–174 (2009)

11. Karwatzki, S., Dytynko, O., Trenz, M., Veit, D.: Beyond the personalization-privacy paradox: privacy valuation, transparency features, and service personalization. J. Manag. Inf. Syst. **34**, 369–400 (2017)

12. Smith, H.J., Dinev, T., Xu, H.: Information privacy research: an interdisciplinary review. MIS Q. **35**, 989–1015 (2011)

13. Li, Y.: The impact of disposition to privacy, website reputation and website familiarity on information privacy concerns. Decis. Support Syst. **57**, 343–354 (2014)

14. Dinev, T., McConnell, A.R., Smith, H.J.: Research commentary - informing privacy research through information systems, psychology, and behavioral economics: thinking outside the "APCO" box. Inf. Syst. Res. **26**, 639–655 (2015)

15. Bélanger, F., Crossler, R.E.: Privacy in the digital age: a review of information privacy research in information systems. MIS Q. **35**, 1017–1041 (2011)

16. Wottrich, V.M., van Reijmersdal, E.A., Smit, E.G.: The privacy trade-off for mobile app downloads: the roles of app value, intrusiveness, and privacy concerns. Decis. Support Syst. **106**, 44–52 (2018)

17. Sutanto, J., Palme, E., Tan, C.-H., Phang, C.W.: Addressing the personalization-privacy paradox: an empirical assessment from a field experiment on smartphone users. MIS Q. **37**, 1141–1164 (2013)

18. Treiblmaier, H., Pollach, I.: Users' perceptions of benefits and costs of personalization. In: ICIS (2007)

19. Xu, H., Teo, H.-H., Tan, B.C.Y., Agarwal, R.: Research note-effects of individual self-protection, industry self-regulation, and government regulation on privacy concerns: a study of location-based services. Inf. Syst. Res. **23**, 1342–1363 (2012)

20. Mousavizadeh, M., Kim, D.J., Chen, R.: Effects of assurance mechanisms and consumer concerns on online purchase decisions: an empirical study. Decis. Support Syst. **92**, 79–90 (2016)

21. Min, J., Kim, B.: How are people enticed to disclose personal information despite privacy concerns in social network sites? The calculus between benefit and cost. J. Assoc. Inf. Sci. Technol. **66**, 839–857 (2015)

22. Dienlin, T., Metzger, M.J.: An extended privacy calculus model for SNSs: analyzing self-disclosure and self-withdrawal in a representative US sample. J. Comput. Commun. **21**, 368–383 (2016)

23. Hui, K.-L., Teo, H.H., Lee, S.-Y.T.: The value of privacy assurance: an exploratory field experiment. MIS Q. **31**, 19–33 (2007)

24. Anderson, C.L., Agarwal, R., Anderson, C.L.: The digitization of healthcare: boundary risks, emotion, and consumer willingness to disclose personal health information. Inf. Syst. Res. **22**, 469–490 (2011)

25. Lee, H., Lim, D., Kim, H., Zo, H., Ciganek, A.P.: Compensation paradox: the influence of monetary rewards on user behaviour. Behav. Inf. Technol. **34**, 45–56 (2015)

26. Jensen, M.L., Yetgin, E.: Prominence and interpretation of online conflict of interest disclosure. MIS Q. **41**, 629–643 (2017)

27. Fogg, B.J.: Prominence-interpretation theory: explaining how people assess credibility online. In: Conference Human Factors Computing Systems – Proceedings, pp. 722–723 (2003)

28. Beatty, M.J., Behnke, R.R.: Teacher credibility as a function of verbal content and paralinguistic cues. Commun. Q. **28**, 55–59 (1980)

29. Beatty, M.J., Kruger, M.W.: The effects of heckling on speaker credibility and attitude change. Commun. Q. **28**, 46–50 (1978)

30. Ohanian, R.: Construction and validation of a scale to measure celebrity endorsers' perceived expertise, trustworthiness, and attractiveness. J. Advert. **19**, 39–52 (1990)

31. Fogg, B.J., et al.: What makes web sites credible?: a report on a large quantitative study. In: Proceedings of the SIGCHI Conference on Human Factors in Computing Systems, pp. 61–68. ACM, New York, NY, USA (2001)

32. Spry, A., Pappu, R., Bettina Cornwell, T.: Celebrity endorsement, brand credibility and brand equity. Eur. J. Mark. **45**, 882–909 (2011)

33. Siehl, C., Bowen, D.E., Pearson, C.M.: Service encounters as rites of integration: an information processing model. Organ. Sci. **3**, 537–555 (1992)

34. George, J.F., Giordano, G., Tilley, P.A.: website credibility and deceiver credibility: expanding prominence-interpretation theory. Comput. Human Behav. **54**, 83–93 (2016)

35. Gefen, D., Rigdon, E.E., Straub, D.: Editor's Comments: an update and extension to SEM guidelines for administrative and social science research. MIS Q. **35**, iii–xiv (2011)

36. McKnight, D.H., Choudhury, V., Kacmar, C.: Developing and validating trust measures for e-commerce: an integrative typology. Inf. Syst. Res. **13**, 334–359 (2002)

37. Reimer, T., Benkenstein, M.: When good WOM hurts and bad WOM gains: the effect of untrustworthy online reviews. J. Bus. Res. **69**, 5993–6001 (2016)

38. Martin, W.C., Lueg, J.E.: Modeling word-of-mouth usage. J. Bus. Res. **66**, 801–808 (2013)

39. Mullins, R.R., Bachrach, D.G., Rapp, A.A., Grewal, D., Beitelspacher, L.S.: You don't always get what you want, and you don't always want what you get: an examination of control–desire for control congruence in transactional relationships. J. Appl. Psychol. **100**, 1073 (2015)

40. Cook, J., Wall, T.: New work attitude measures of trust, organizational commitment and personal need non-fulfilment. J. Occup. Psychol. **53**, 39–52 (1980)

41. Cialdini, R.B.: Influence: science and practice. Pearson education, Boston (2009)

42. Fogg, B.J., Tseng, H.: The elements of computer credibility. In: Proceedings of the SIGCHI Conference on Human Factors in Computing Systems. pp. 80–87. ACM, New York, NY, USA (1999)

43. Mayer, R.C., Davis, J.H.: The effect of the performance appraisal system on trust for management: a field quasi-experiment. J. Appl. Psychol. **84**, 123 (1999)

44. Kelley, H.H., Michela, J.L.: Attribution theory and research. Annu. Rev. Psychol. **31**, 457–501 (1980)

45. Whiting, S.W., Maynes, T.D., Podsakoff, N.P., Podsakoff, P.M.: Effects of message, source, and context on evaluations of employee voice behavior. J. Appl. Psychol. **97**, 159 (2012)

46. Munnukka, J., Uusitalo, O., Toivonen, H.: Credibility of a peer endorser and advertising effectiveness. J. Consum. Mark. **33**, 182–192 (2016)

47. Bansal, G., Zahedi, F.M., Gefen, D.: The role of privacy assurance mechanisms in building trust and the moderating role of privacy concern. Eur. J. Inf. Syst. **24**, 624–644 (2015)

48. Yu, J., Hu, P.J.H., Cheng, T.H.: Role of affect in self-disclosure on social network websites: a test of two competing models. J. Manag. Inf. Syst. **32**, 239–277 (2015)

e-Market

Competitive Analysis of "Buy Online and Pick Up in Store" Channel

Ronghui Wang, Lin Chen, Haiyang Feng, Guofang Nan$^{(\boxtimes)}$, and Minqiang Li

College of Management and Economics, Tianjin University,
Tianjin 300072, China
{ronghuiwang, linchen, hyfeng, gfnan, mqli}@tju.edu.cn

Abstract. In this paper, we model the competition between two firms that each firm can choose a "buy online and pick up in store" (BOPS) channel strategy or an offline channel strategy. We find out when it is optimal for a retailer to offer a BOPS channel by comparing the equilibrium outcomes. Whether to adopt the BOPS channel depends largely on the difference between the inherent values provided by two firms. Further, we prove that a prisoner's dilemma exists under certain condition when two competing firms offer the BOPS channel. We also examine the impact of heterogeneous channel acceptance on the profits of two firms. We find that when only a firm offers the BOPS channel, the higher acceptance of BOPS means the higher profit of this firm and the less profit of its competitor. However, when both firms offer the BOPS channel, whether a firm can benefit from the increase in acceptance of BOPS depends on the inherent value it provides.

Keywords: E-commerce · Buy online and pick up in store ·
Omnichannel retailing · Competitive strategy · Pricing

1 Introduction

The advent of the e-commerce era has significantly changed the way consumers shop [1]. As offline retailers realize the advantages of online channels, they begin to offer multiple channels and conduct cross-channel integration, thus, promoting the rise of omnichannel retailing, which carries out cross-channel integration in all available shopping channels to meet the shopping needs of customers in different scenarios [2]. According to the 2018 report compiled by Total Retail and Radial, 52% of omnichannel retailers among the top 100 offer the "buy online and pick up in store" (BOPS) channel. Although the BOPS channel is relatively new for some consumers, it has been widely adopted in many industries, such as the clothing, catering, and retailing industries.

From the consumer's perspective, the BOPS channel has both advantages and disadvantages when compared to the offline channel. To some extent, the BOPS channel has combined some superior qualities of the online and offline channels, and can offer a higher convenience level. For example, consumers can easily get a greater variety of products through the BOPS channel, than through the offline channel,

© Springer Nature Switzerland AG 2019
J. J. Xu et al. (Eds.): WEB 2018, LNBIP 357, pp. 65–77, 2019.
https://doi.org/10.1007/978-3-030-22784-5_7

because the physical store has limited space. In addition, the BOPS channel provides consumers with more conveniences when they visit to pick up their purchases, such as dedicated parking lanes and quick cashier lines [3]. All the above characteristics of the BOPS channel enable consumers to enjoy a more convenient shopping experience. However, the BOPS channel is not always superior to the offline channel. If a consumer finds that a product he purchased does not fit him when he picks it up in the store, he cannot get a full refund right away. Considering that the BOPS order is prepaid online, it usually takes a few business days to return the refund to the original payment account, which mainly depends on the bank's processing time. Besides, some retailers may require consumers who choose the BOPS channel to pay an extra fee, such as, Kmart and IKEA in Australia. In some instances, certain new releases are only sold through the offline channel, instead of the BOPS channel. For example, in Toys "R" Us, new releases are not available for the BOPS channel [2].

Our paper considers the different strategic combinations generated by two firms' choices between the offline channel and the BOPS channel. We specifically concentrate on answering the following three research questions. First, under what conditions will the firm be better off offering the BOPS channel rather than offering the offline channel? Second, what factors affect the price and channel convenience level of a firm and how? Finally, in different sub-games, are the payoffs in the equilibrium of two firms optimal? If external factors change, what will the impact on the profits of two competing firms be?

To analyze the above questions, we set up a model in which two competing firms sell differentiated products, and each firm can offer the BOPS channel to provide their consumers with a more efficient shopping experience. We take heterogeneous channel acceptance, consumer convenience sensitivity coefficient, and cost coefficient into consideration. We arrive at a few valuable conclusions on whether to offer the BOPS channel in the competition environment or not. First, we show that a firm's strategy concerning the decision to provide the BOPS channel depends on the difference between the inherent value that consumers obtain from two competing firms, and sometimes, also on its competitor's strategy. When the difference in inherent value for purchases from two firms is not obvious, choosing the same strategy (i.e., BOPS channel or offline channel) is the best option for both firms. In this scenario, a firm's channel strategy is mainly affected by the competitor's channel strategy. However, when a firm has a significant advantage in inherent value for a purchase, a unique equilibrium prevails. Second, we find a prisoner's dilemma for the two firms choosing the BOPS channel when they offer similar inherent value to consumers. In this scenario, although the firm's original intention is to increase demand and improve market competitiveness by adopting the BOPS channel, the extra cost of adopting the BOPS channel has a significant negative impact on its profit. However, when the purchases are differentiated enough in the inherent value they provide to consumers, the firm that offers lower inherent value would be better off when both firms sell products through the BOPS channel.

The rest of our paper is structured as follows. Section 2 reviews the literature. In Sect. 3, we first introduce the setup of the model and then show the optimal prices and optimal channel convenience levels of the two firms in the four sub-games. Section 4 compares the equilibrium outcomes in relation to the two firms' channel strategies. Section 5 concludes the paper.

2 Literature Review

Our paper is related to two streams of work: retailing channel choice in the omnichannel environment and competition strategy in a duopoly.

With the booming development of e-commerce, channel integration has been regarded as a promising strategy for retailers. In this context, omnichannel retailing has received considerable attention in academia [4]. Many studies address ways to serve consumers better in the omnichannel retailing industry. Gallino et al. [5] empirically test the impact of introducing the ship to store option (one of the most widely used omnichannel fulfillment options) on the retailer's sales dispersion and investigate the effect of channel integration on inventory management. Gao and Su [6] study three information mechanisms including physical showrooms, virtual showrooms, and availability information, and they focus on the efficiency of online and offline information transferring from retailers to consumers who face product value uncertainty and availability uncertainty in an omnichannel environment.

The stream of work on the retailing channel choice in the omnichannel environment has studied the BOPS channel, from both, empirical and theoretical perspectives. Gallino and Moreno [7] conduct an empirical test on the effect of the BOPS channel on the sales in both online and store channels of a retailer. They show that if the BOPS channel is adopted, then the online sales will decrease and offline sales will increase. In addition, there are some theoretical models in this area as well [3, 8]. Cao et al. [3] investigate the impact of the BOPS channel on demand allocation and the profitability of a retailer adopting multiple channels to sell its products. They find that not all products are suitable for sale through the BOPS channel. Jin et al. [9] establish a theoretical model where a store retailer offering the BOPS channel fulfills online and offline orders through a recommended service area and they find that the unit inventory cost and the consumers' arrival rate of BOPS channel play a decisive role in the scale of service area. Gao and Su [10] focus on the application of self-service technologies including online and offline technologies in the restaurant industry and analyze the impact of self-order technologies on the demand and profits of a restaurant. While most studies in this area consider the case of a retailer with multi-channels and examine the effect of the BOPS channel on the retailer's profits, our paper analyzes competition between two retailers in a game theoretical framework to determine the optimal decisions of the two retailers in different situations. Our paper also considers heterogeneous channels and investigates their impact on the firms' profits. Moreover, considering the variable consumer convenience sensitivity coefficient, we set the endogenous channel convenience level and introduce a quadratic function of the additional cost related to the channel convenience level. Furthermore, we analyze the optimal strategies for both firms under different conditions and find a prisoner's dilemma.

Our paper is also relevant to the work on the competitive strategies of two firms. In this stream, a majority of studies adopt the game theory to analyze the price competition in a duopoly. Etzion and Pang [11] investigate the competition between two firms selling a differentiated physical product and each firm can choose to provide a complementary online service that exhibits network effects and is independent of the product. They find that while network effects intensify price competition, the growth of the network effect benefits the firm. In the supply chain setting, Chen and Guo [12] focus on the case where a common supplier provides two downstream retailers with uncertain supply and investigate the price competition between two retailers in a horizontally differentiated product market. Jena and Sarmah [13] investigate the co-operation problems and price competition in a closed-loop supply chain with two competitive manufacturers and a common retailer. Different from the above studies, our paper not only examines the case where the price competition exists between two firms but also studies the case where price competition and convenience level competition co-exist. In particular, when both firms offer the BOPS channel, competition for convenience level is added to the existing competition between the two firms. In addition to the above on price competition, there is some related literature considering non-price competition. Xiao et al. [14] study strategic outsourcing decisions in the context of two manufacturers competing on the quality of products. Ding et al. [15] consider the inventory factor and environmental constraints to investigate the service competition between two retailers in an online market. Zhao et al. [16] focus on quality disclosure strategies for two small business enterprises (SBEs) in a competitive supply chain setting, and show that the information asymmetry does have an effect on an SBE's strategy on whether to disclose the quality to customers through the retailer or not. The above studies all think that the price is not the only determinant of consumer decisions, most of which are concerned with quality competition in a duopoly. However, the convenience level provided by a firm is also playing a crucial role in consumers' purchase choice, which remains to be studied. In this paper, we focus on the competitive model in which there are two firms that choose to provide the BOPS channel or a pure offline channel according to the situation. Further, we consider that the two firms compete not only on price, but also on channel convenience level. We investigate four possible pairs of channel strategies with regard to the sub-game of both firms adopting the BOPS channel, the two sub-games of only one firm adopting the BOPS channel, and the sub-game of neither firm adopting the BOPS channel.

3 Model

Consider two firms selling differentiated products in a common market. Each firm chooses to sell through the offline channel or offers the product to the customer via the novel BOPS channel. Since we only consider the competition between two firms in one market and the number of firms is fixed without considering new entrants, it is appropriate to adopt the Hotelling setup [17]. We assume that the product of Firm 1 is located at 0 and the product of Firm 2 is located at 1 in our model. Considering the different travel costs for consumers going to the store, consumers in the market are heterogeneous in the matter of their product preferences. When a consumer located at point x buys from Firm 1, the travel cost can be denoted by tx, which is proportionate to the distance between the consumer and the product, and when the consumer located at point x buys from Firm 2, the travel cost is $t(1-x)$. Without loss of generality, we set $t = 1$, where t means the per-unit travel cost [18, 19].

For consumers, the offline channel and the BOPS channel are different in two aspects: convenience level and channel acceptance. We denote convenience levels for the BOPS channel and offline channel by s and c, respectively. Herein, we assume that $s > c$. Without loss of generality, we set $c = 0$ and make s represent the extra convenience of the BOPS channel when compared to the offline channel. The extra utility a consumer obtains from the BOPS channel is given by rs, where r is convenience sensitivity and $0 < r < 1$. The more sensitive consumers are to convenience, the more utility they can get from the BOPS channel.

Compared with the offline channel, the BOPS channel has advantages in terms of channel convenience level, but it is still insufficient in terms of channel acceptance. It is common sense that the offline channel is more mature and has accumulated a large consumer base. In contrast, the BOPS channel has some immature phenomena now, such as unreasonable store recommendation [9]. Therefore, it is reasonable to assume that the inherent value for a purchase from Firm i when Firm i chooses the offline channel is $v_i (i = 1, 2)$ which depends on the inherent value of the product and on the variety of products in a firm, while the value for a purchase from Firm i when Firm i chooses the BOPS channel is $\theta v_i (i = 1, 2)$, where $0 < \theta < 1$.

When a firm offers the BOPS channel, there will be an additional cost due to the higher convenience level. In this model, we assume that the additional cost is a quadratic function and is denoted by $\frac{ks^2}{2}$, where k represents the cost coefficient of improving the convenience level and when the value of k decreases, the cost efficiency increases. The quadratic cost function is very common in previous literatures [20, 21]. What's more, the quadratic functional form for the additional cost is the most tractable way of getting an analytic solution in our model.

Based on the above model descriptions, we can summarize the consumer utility functions as follows:

$$U_O^1 = v_1 - p_g^1 - x, \tag{1}$$

$$U_B^1 = \theta v_1 - p_g^1 - x + rs_g^1, \tag{2}$$

$$U_O^2 = v_2 - p_g^2 - (1 - x), \tag{3}$$

$$U_B^2 = \theta v_2 - p_g^2 - (1 - x) + rs_g^2, \tag{4}$$

where Eq. (1) is the consumer utility when a consumer buys from the offline channel provided by Firm 1, Eq. (2) is the consumer utility when a consumer buys from the BOPS channel provided by Firm 1, Eq. (3) is the consumer utility when a consumer buys from the offline channel provided by Firm 2, and Eq. (4) is the consumer utility when a consumer buys from the BOPS channel provided by Firm 2.

There can be four pairs of channel strategies: both firms offer the offline channel, marked as (O, O), Firm 1 chooses the offline channel while Firm 2 chooses the BOPS channel, marked as (O, B), Firm 1 chooses the BOPS channel while Firm 2 chooses the offline channel, marked as (B, O), and both firms offer the BOPS channel, marked as (B, B).

3.1   Sub-Game (O,O): Competition Between Two Firms Offering the Pure Offline Channel

Under sub-game (O,O), both firms choose to provide the pure offline channel, and all the consumers go shopping through the offline channel of Firm 1 or Firm 2. Considering that different consumers have different preferences for the two firms, the market can be divided into two segments by the marginal consumer whose location is denoted as x_0. By setting consumer utility function $U_O^1 = U_O^2$, we can derive the point of indifference $x_0 = \frac{v_1 - v_2 - p_{OO}^1 + p_{OO}^2 + 1}{2}$, where the marginal consumer is indifferent to purchase from the channel provided by Firm 1 or Firm 2.

In this sub-game, the two firms need to set their prices p_{OO}^1 and p_{OO}^2 to maximize their profit without considering the convenience level because the convenience level of each firm is equal to zero. The profit functions of the two firms offering the offline channel are as follows:

$$\pi_{OO}^1 = p_{OO}^1 D_{OO}^1 = p_{OO}^1 x_0.$$
$$\pi_{OO}^2 = p_{OO}^2 D_{OO}^2 = p_{OO}^2 (1 - x_0), \tag{5}$$
$$s.t. 0 < x_0 < 1.$$

Substituting x_0 into both the profit functions mentioned above, we can get the equilibrium solutions for this sub-game. Solving the first order conditions of the two profit functions at the same time, we derive the following optimal prices p_{OO}^{1*} and p_{OO}^{2*} when both firms offer the offline channel:

$$p_{OO}^{1*} = \frac{v_1 - v_2 + 3}{3}, p_{OO}^{2*} = \frac{v_2 - v_1 + 3}{3} \tag{6}$$

According to the optimal prices, we derive the product demands of the two firms and then get their profits as follows:

$$D_{OO}^{1*} = \frac{v_1 - v_2 + 3}{6}, D_{OO}^{2*} = \frac{v_2 - v_1 + 3}{6}, \pi_{OO}^{1*} = \frac{(v_1 - v_2 + 3)^2}{18}, \pi_{OO}^{2*} = \frac{(v_2 - v_1 + 3)^2}{18}$$
$$\tag{7}$$

3.2   Sub-Game (O,B)/Sub-Game (B,O): Competition When Only One Firm Offers the BOPS Channel

Under sub-game (O,B) or sub-game (B,O), only one firm offers the BOPS channel while its competitor offers the offline channel. Without loss of generality, we consider that Firm 1 chooses to offer the offline channel while Firm 2 chooses to offer the BOPS channel, since the sub-game (B,O) is symmetrically identical.

In this sub-game, consumers who are indifferent to the purchase channels, located at x_1, can divide the market into two segments. Consumers locating on the left of x_1 prefer to purchase from the offline channel of Firm 1 while consumers located on the right of x_1 prefer to purchase from the BOPS channel of Firm 2. Solving the indifference equation $U_O^1 = U_B^2$, we derive the point of indifference thus:

$$x_1 = \frac{v_1 - \theta v_2 - p_{OB}^1 + p_{OB}^2 + 1 - rs_{OB}^2}{2}. \tag{8}$$

In equilibrium, Firm 1 sets the price p_{OB}^1 to maximize its profit from the offline channel, while Firm 2 sets the price p_{OB}^2 and the convenience level s_{OB}^2 to maximize its profit from the BOPS channel. The profit functions for the two firms are as follows:

$$\pi_{OB}^1 = p_{OB}^1 D_{OB}^1 = p_{OB}^1 x_1,$$

$$\pi_{OB}^2 = p_{OB}^2 D_{OB}^2 - \frac{k(s_{OB}^2)^2}{2} = p_{OB}^2(1 - x_1) - \frac{k(s_{OB}^2)^2}{2}, \tag{9}$$

$$s.t. 0 < x_1 < 1.$$

There is no doubt that the profit function of Firm 1 is concave in p_{OB}^1. To ensure the concavity of the profit function of Firm 2, we get a constraint $4k > r^2$ by solving for the Hessian matrix of π_{OB}^2. Therefore, we can get the optimal convenience level of Firm 2 and the optimal prices of the two firms:

$$s_{OB}^{2*} = \frac{r(\theta v_2 - v_1 + 3)}{6k - r^2}, p_{OB}^{1*} = \frac{2(kv_1 - k\theta v_2 + 3k - r^2)}{6k - r^2}, p_{OB}^{2*} = \frac{2k(\theta v_2 - v_1 + 3)}{6k - r^2}. \tag{10}$$

Based on the optimal convenience level and the optimal prices, we can derive the demand of Firm 1's offline channel and the demand of Firm 2's BOPS channel:

$$D_{OB}^{1*} = \frac{kv_1 - k\theta v_2 + 3k - r^2}{6k - r^2}, D_{OB}^{2*} = \frac{k(\theta v_2 - v_1 + 3)}{6k - r^2}. \tag{11}$$

Therefore, the profits of the two firms are as follows:

$$\pi_{OB}^{1*} = \frac{2(kv_1 - k\theta v_2 + 3k - r^2)^2}{(6k - r^2)^2}, \pi_{OB}^{2*} = \frac{k(\theta v_2 - v_1 + 3)^2(4k - r^2)}{2(6k - r^2)^2}. \tag{12}$$

3.3 Sub-Game (B,B): Competition Between Two Firms Offering the BOPS Channel

Under sub-game (B, B), the two firms choose to offer the BOPS channel with different convenience levels. Let x_2 represent the position of the marginal customer. Consumers can be divided into two segments: those who are located on the left of x_2 choose to buy from the BOPS channel of Firm 1 and consumers who are located on the right of x_2 choose to buy from the BOPS channel of Firm 2. By setting consumer utility function

$U_B^1 = U_B^2$, we work out the point of indifference and the customers here are indifferent about purchasing from Firm 1 or Firm 2.

$$x_2 = \frac{\theta v_1 - \theta v_2 - p_{BB}^1 + p_{BB}^2 + 1 + rs_{BB}^1 - rs_{BB}^2}{2}. \tag{13}$$

Given the consumers' choices, the two firms determine the different convenience levels for the channel and the prices for the products to obtain their maximum profits. The profit functions when both firms offer the BOPS channel are as follows:

$$\pi_{BB}^1 = p_{BB}^1 D_{BB}^1 - \frac{k(s_{BB}^1)^2}{2} = p_{BB}^1 x_2 - \frac{k(s_{BB}^1)^2}{2},$$

$$\pi_{BB}^2 = p_{BB}^2 D_{BB}^2 - \frac{k(s_{BB}^2)^2}{2} = p_{BB}^2 (1 - x_2) - \frac{k(s_{BB}^2)^2}{2}, \tag{14}$$

$$s.t. 0 < x_2 < 1.$$

We derive a constraint that guarantees the concavity of the profit functions, which is given by $4k > r^2$. Solving $\frac{\partial \pi_{BB}^1}{\partial s_{BB}^1} = 0, \frac{\partial \pi_{BB}^1}{\partial p_{BB}^1} = 0, \frac{\partial \pi_{BB}^2}{\partial s_{BB}^2} = 0$, and $\frac{\partial \pi_{BB}^2}{\partial p_{BB}^2} = 0$ simultaneously, we get the optimal convenience levels and the optimal prices of the two firms:

$$s_{BB}^{1*} = \frac{r(k\theta v_1 - k\theta v_2 + 3k - r^2)}{2k(3k - r^2)}, s_{BB}^{2*} = \frac{r(k\theta v_2 - k\theta v_1 + 3k - r^2)}{2k(3k - r^2)}. \tag{15}$$

$$p_{BB}^{1*} = \frac{k\theta v_1 - k\theta v_2 + 3k - r^2}{3k - r^2}, p_{BB}^{2*} = \frac{k\theta v_2 - k\theta v_1 + 3k - r^2}{3k - r^2} \tag{16}$$

From the above optimal convenience levels and the optimal prices, we can obtain the demands of the BOPS channels of both Firm 1 and Firm 2 as follows:

$$D_{BB}^{1*} = \frac{k\theta v_1 - k\theta v_2 + 3k - r^2}{2(3k - r^2)}, D_{BB}^{2*} = \frac{k\theta v_2 - k\theta v_1 + 3k - r^2}{2(3k - r^2)}. \tag{17}$$

Accordingly, the profit functions of the two firms are as follows:

$$\pi_{BB}^{1*} = \frac{(k\theta v_1 - k\theta v_2 + 3k - r^2)^2 (4k - r^2)}{8k(3k - r^2)^2}, \pi_{BB}^{2*} = \frac{(k\theta v_2 - k\theta v_1 + 3k - r^2)^2 (4k - r^2)}{8k(3k - r^2)^2}. \tag{18}$$

4 Market Equilibrium Analysis

In the previous section, by adopting the concept of sub-game perfectness and solving the three-stage game backward, we derive the optimal convenience levels, the optimal prices, and the optimal profits of the two firms associated with the four possible market configurations. Based on the results shown above, we can obtain the following market equilibrium.

Proposition 1 (Market equilibrium under different conditions):

(1) Sub-game (O,O) is an equilibrium in which both firms offer the offline channel, if and only if $v_1 > A_1 v_2 + 3$ and $v_2 > A_1 v_1 + 3$.

(2) Sub-game (B,O) is an equilibrium in which only firm 1 offers the BOPS channel, if and only if $v_1 > A_2 v_2 + \frac{3k - r^2}{k\theta}$ and $v_2 < A_1 v_1 + 3$.

(3) Sub-game (O,B) is an equilibrium in which only firm 2 offers the BOPS channel, if and only if $v_1 < A_1 v_2 + 3$ and $v_2 > A_2 v_1 + \frac{3k - r^2}{k\theta}$.

(4) Sub-game (B,B) is an equilibrium in which both firms offer the BOPS channel, if and only if $v_1 < A_2 v_2 + \frac{3k - r^2}{k\theta}$ and $v_2 < A_2 v_1 + \frac{3k - r^2}{k\theta}$.

Where $A_1 = \frac{3\theta\sqrt{k(4k - r^2)} - 6k + r^2}{3\sqrt{k(4k - r^2)} - 6k + r^2}$, $A_2 = \frac{\theta(6k - r^2)\sqrt{4k - r^2} - 4\sqrt{k}(3k - r^2)}{\theta(6k - r^2)\sqrt{4k - r^2} - 4\theta\sqrt{k}(3k - r^2)}$.

According to Proposition 1, we can derive some meaningful strategic implications. When a firm takes a decision on whether to provide the BOPS channel, the firm needs to take into consideration the inherent value that consumers can get while buying from its competitor. The variable market factors could also influence the firm's channel strategy, including the channel acceptance θ, convenience sensitivity r, and cost coefficient k. In order to provide a more intuitive description of Proposition 1, we depict the market equilibrium outcomes under different conditions in Fig. 1, with v_1 on the horizontal axis and v_2 on the vertical axis. For the sake of simplicity, we adopt the italic capitals (O,O), (O,B), (B,O), and (B,B) to represent the corresponding four sub-games.

From Fig. 1 we can find that if a firm has a significant advantage in inherent value, it is optimal for the firm to choose to provide consumers with the BOPS channel while for the competitor the offline channel is optimal. When the inherent value from Firm 1 is apparently higher than the inherent value from Firm 2, Firm 1 offering the BOPS channel and Firm 2 offering the offline channel could both obtain more profits than adopting any other strategy combination, as shown in region f-n-i. Symmetrically, when the inherent value for a purchase from Firm 2 is much higher than the inherent value for a purchase from Firm 1, it is more profitable for Firm 2 to offer the BOPS channel and for Firm 1 to offer the offline channel, as shown in region d-m-g.

In addition, if the inherent value for a purchase provided by the two firms is not significantly different, there would be two situations. When both firms bring low inherent value to consumers, the state of equilibrium is one where both firms choose to sell through the BOPS channel, as shown in region d-m-e-n-f. In general, for a firm that provides low inherent value for consumers, if the firm sells through the physical store, the cost of developing the offline channel will affect the total profits greatly.

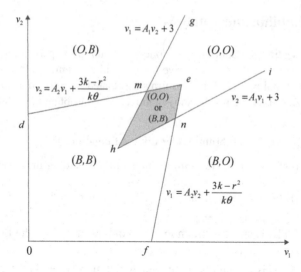

Fig. 1. Market equilibrium

In contrast, when both firms can bring considerable inherent value to consumers, it is optimal for both firms to adopt the offline channel, as shown in region g-m-h-n-i. That is, if the inherent value of the two firms is both high enough, each firm would choose the offline channel, regardless of its competitor's strategy. To a certain extent, the high inherent value for a purchase from the firm can generate enough attraction for consumers, so there is no need for the firm to employ the BOPS channel to attract consumers.

For a set of the parameter values, as shown in the region h-m-e-n, two feasible equilibriums exist. One is where both firms adopt the offline channel, and the other is where both firms adopt the BOPS channel. In other words, each firm finds it profitable to adopt the same channel strategy as its competitor within this range of parameter values.

After analyzing the equilibrium outcomes of the game, sometimes, we find that the equilibrium is not the Pareto optimal. That is, we find that the two firms may be caught in a prisoner's dilemma in some situations, as stated in Proposition 2.

Proposition 2 (Prisoner's dilemma condition): When both firms choose the BOPS channel, Firm i would be better off when neither chooses it if and only if

$$v_j - v_i < \frac{\left(6\sqrt{k} - 3\sqrt{4k-r^2}\right)\left(3k-r^2\right)}{2\sqrt{k}(3k-r^2) - 3k\theta\sqrt{4k-r^2}}.$$

According to Proposition 2, we find that if the two firms choose to sell through the BOPS channel when the difference in terms of the inherent value for a purchase between two firms is not significant, both firms would get more profits through the offline channel rather than the BOPS channel. That is, the two firms would be caught in a prisoner's dilemma because of the intensified competition when $|v_j - v_i|$ is small enough. On this occasion, both firms are likely to adopt the BOPS strategy to increase the demands and compete for the market, but the extra cost of adopting the BOPS channel can have a significant negative impact on their profits. Nevertheless, if Firm i has a significant disadvantage in the inherent value for a purchase over Firm j, then Firm i obtains more profits when both firms adopt the BOPS channel to sell their products.

Now, we analyze how the profit of a firm changes with an increase in channel acceptance of BOPS θ and the results are shown in Proposition 3.

Proposition 3 (Impact of heterogeneous channel acceptance on profits):

(1) When only Firm i offers the BOPS channel in equilibrium, it always benefits from an increase in θ. An increase in θ has a negative effect on Firm j's profit.

(2) When both firms offer the BOPS channel, for Firm i and Firm j, an increase in θ has a positive effect on Firm i's profit and a negative effect on Firm j's profit if the inherent value of Firm i is higher than the inherent value of Firm j (i.e., $v_i > v_j$).

Proposition 3 shows that if only one firm adopts the BOPS channel, the firm's profit will continue to rise as consumers become more acceptable to the BOPS channel. However, with the increase in the channel acceptance, more consumers are willing to buy products from the BOPS channel, thus leading to the decrease in competing firm's profit earned from the offline channel.

On the other hand, when both firms choose to sell their products through the BOPS channel, the profits of both firms do not necessarily increase in the channel acceptance due to the difference in the inherent value provided by the two firms. The firm can obtain more profit from an increase in channel acceptance only if it can bring consumers more inherent value than its competitor. However, the increase in Firm j's demand may not be enough to offset the lost profit due to the price reduction and the extra cost for offering the BOPS channel, which explains why the profit of the firm, whose inherent value for a purchase is not dominant, will decrease with the increase of channel acceptance.

5 Conclusion

In this paper, we investigate whether two competing firms offer the BOPS channel by considering four pairs of channel strategies, and if they do, then when. By analyzing the equilibrium outcomes, we find that whether a firm adopts the BOPS channel or not depends largely on the difference between the inherent values that consumers obtain from each of the two firms. In particular, when one firm provides an obviously higher inherent value over its competitor, adopting the BOPS channel is the optimal option for this firm, but for the other firm, the offline channel should be chosen. Furthermore, we find that competing firms are caught in a prisoner's dilemma when they offer the BOPS channel under certain conditions.

Our work has several implications for managers who make decisions on whether to provide the offline channel or the BOPS channel in a duopoly. First, when the inherent value gap for a purchase between the two firms is not particularly large, if both firms provide high enough inherent value, they should adopt the offline channel strategy. In this scenario, offline sales are attractive enough for consumers, that there is no need for firms to adopt the BOPS channel. Second, if a firm has an obvious advantage in inherent value for a purchase, then the firm should offer the BOPS channel and its competitor should offer the offline channel. The firm with an inherent value advantage will have a dual appeal to consumers by using the BOPS channel to improve its convenience level. However, for the other firm, adopting the BOPS channel means incurring a high cost and earning little profit. Thus, providing an offline channel is the best choice.

This paper can be extended in the following directions. First, to simplify the theoretical model, some factors that influence the popularity of the BOPS channel, including the strict return policy, the extra fee for BOPS consumers, and the unreasonable store recommendation, are indirectly reflected in the parameter of heterogeneous channel acceptance. However, if these factors are shown directly in the model, there can be some instructive discoveries. Second, in the current work, we analyze the competition between two firms in the context of each firm adopts only one channel. However, what we observe in practice are mostly hybrid channels, that is, many firms actually sell products through multiple channels, such as the online channel, the offline channel, the BOPS channel and so on. Taking into account all the channels offered by both firms may be an interesting direction.

Acknowledgments. This research was partially supported by research grant from the National Science Foundation of China (No. 71471128) and the Key Program of National Natural Science Foundation of China (No. 71631003).

References

1. Feng, Y., Guo, Z., Chiang, W.K.: Optimal digital content distribution strategy in the presence of the consumer-to-consumer channel. J. Manage. Inform. Syst. **25**, 241–270 (2009). https://doi.org/10.2753/mis0742-1222250408
2. Gao, F., Su, X.: Omnichannel retail operations with buy-online-and-pick-up-in-store. Manage. Sci. **63**, 2478–2492 (2017). https://doi.org/10.1287/mnsc.2016.2473
3. Cao, J., So, K.C., Yin, S.: Impact of an "online-to-store" channel on demand allocation, pricing and profitability. Eur. J. Oper. Res. **248**, 234–245 (2016). https://doi.org/10.1016/j.ejor.2015.07.014
4. Li, Y., Liu, H., Lim, E.T.K., Goh, J.M., Yang, F., Lee, M.K.O.: Customer's reaction to cross-channel integration in omnichannel retailing: the mediating roles of retailer uncertainty, identity attractiveness, and switching costs. Decis. Support Syst. **109**, 50–60 (2018). https://doi.org/10.1016/j.dss.2017.12.010
5. Gallino, S., Moreno, A., Stamatopoulos, I.: Channel integration, sales dispersion, and inventory management. Manage. Sci. **63**, 2813–2831 (2017). https://doi.org/10.1287/mnsc.2016.2479
6. Gao, F., Su, X.: Online and offline information for omnichannel retailing. Manuf. Serv. Oper. Manag. **19**, 84–98 (2017). https://doi.org/10.1287/msom.2016.0593
7. Gallino, S., Moreno, A.: Integration of online and offline channels in retail: the impact of sharing reliable inventory availability information. Manage. Sci. **60**, 1434–1451 (2014). https://doi.org/10.1287/mnsc.2014.1951
8. Liu, Y., Zhou, D., Chen, X.: Channel integration of BOPS considering off-line sales effort differences. J. Syst. Eng. **33**, 90–102 (2018). https://doi.org/10.13383/j.cnki.jse.2018.01.009

9. Jin, M., Li, G., Cheng, T.C.E.: Buy online and pick up in-store: design of the service area. Eur. J. Oper. Res. **268**, 613–623 (2018). https://doi.org/10.1016/j.ejor.2018.02.002

10. Gao, F., Su, X.: Omnichannel service operations with online and offline self-order technologies. Manage. Sci. **64**, 3595–3608 (2017). https://doi.org/10.1287/mnsc.2017.2787

11. Etzion, H., Pang, M.S.: Complementary online services in competitive markets: maintaining profitability in the presence of network effects. MIS. Q. **38**, 231–248 (2014). https://doi.org/10.25300/MISQ/2014/38.1.11

12. Chen, J., Guo, Z.: Strategic sourcing in the presence of uncertain supply and retail competition. Prod. Oper. Manag. **23**, 1748–1760 (2014). https://doi.org/10.1111/poms.12078

13. Jena, S.K., Sarmah, S.P.: Price competition and co-operation in a duopoly closed-loop supply chain. Int. J. Prod. Econ. **156**, 346–360 (2014). https://doi.org/10.1016/j.ijpe.2014.06.018

14. Xiao, T., Xia, Y., Zhang, G.P.: Strategic outsourcing decisions for manufacturers competing on product quality. IIE Trans. **46**, 313–329 (2014). https://doi.org/10.1080/0740817x.2012.761368

15. Ding, Y., Gao, X., Huang, C., Shu, J., Yang, D.: Service competition in an online duopoly market. Omega **77**, 58–72 (2018). https://doi.org/10.1016/j.omega.2017.05.007

16. Zhao, M., Dong, C., Cheng, T.C.E.: Quality disclosure strategies for small business enterprises in a competitive marketplace. Eur. J. Oper. Res. **270**, 218–229 (2018). https://doi.org/10.1016/j.ejor.2018.03.030

17. Hotelling, H.: Stability in competition. Econ. J. **39**, 41–57 (1929). https://doi.org/10.1007/978-1-4613-8905-7_4

18. Yoo, W.S., Lee, E.: Internet channel entry: a strategic analysis of mixed channel structures. Market. Sci. **30**, 29–41 (2011). https://doi.org/10.1287/mksc.1100.0586

19. Chen, L., Nan, G., Li, M.: Wholesale pricing or agency pricing on retail platforms: the effects of customer loyalty. Int. J. Electron. Comm. **24**, 576–608 (2018). https://doi.org/10.1080/10864415.2018.1485086

20. Iyer, G.: Coordinating channels under price and nonprice competition. Market. Sci. **17**, 338–355 (1998). https://doi.org/10.1287/mksc.17.4.338

21. Moorthy, K.S.: Product and price competition in a duopoly. Mark. Sci. **7**, 141–168 (1988). https://doi.org/10.1287/mksc.7.2.141

An Empirical Analysis of Brand Effects on Online Consumer Repurchase Behavior After Unsatisfied Experience

Xiaojun Luo[(⊠)] and Dan Ke

Wuhan University, Wuhan, China
luo_mimi@qq.com, dkeuconn@gmail.com

Abstract. This paper examines the online repurchase behavior and customer's complaint particularly given the customer has experienced an unsatisfied shopping. We use Least Square Dummy Variable method to construct fixed-effects models and use text mining to analyze Amazon review data. The empirical results show there are brand effects on repurchase behavior after unsatisfied experience. It proves that brand is a crucial factor affecting online consumer repurchase behavior after unsatisfied experience. Transaction-specific satisfaction and overall satisfaction have different influences on repurchase behavior after unsatisfied experience. The opposite effects of these two factors reflect that different kind of satisfaction has different influence on repurchase behavior after unsatisfied experience. We find that after unsatisfied experience, the frequency of repurchase behavior on popular brand is less than that on unpopular brand. Customers prefer to repurchase unpopular brand after unsatisfied experience. This paper figures out that the factors influencing customers' repurchase behavior after unsatisfied experience are more related to product itself, but less to channel.

Keywords: Online repurchase behavior · Unsatisfied experience · Shannon's information theory · Fixed-effects model · Least Square Dummy Variable model

1 Introduction

Consumers' repurchase behavior is a key to a firm. It is one of the most significant factors influencing a firm's financial performance [1] since it ensures that the company's products are sold continuously and profitably. Generally, it is expensive to have lost customers to return, so a firm will do their best to discover customer dissatisfaction and make corresponding efforts to ensure their long-term relationship [2]. Therefore, it is crucial to study the behavior of customers after their unsatisfactory experiences [3].

Nowadays, consumers can not only purchase products through traditional channels, but can in many cases shop online [4, 5]. Logistic regression [3], autoregressive model [6] and SEM models [7] have been used to verify the factors influencing repurchase behavior, but these methods have missing variables, which in turn leads to omitted variable bias. However, there are few studies about customer repurchase behavior after unsatisfied experience. In fact, after customers feel unsatisfied in purchase procedure,

© Springer Nature Switzerland AG 2019
J. J. Xu et al. (Eds.): WEB 2018, LNBIP 357, pp. 78–90, 2019.
https://doi.org/10.1007/978-3-030-22784-5_8

they might still repurchase, which is revealed by online customers' ratings and purchase records. But few researchers consider the condition where customers suffer unsatisfied experience. Thus, whether brand can influence customers repurchase behavior after unsatisfied experience, or what factors can impact such behavior is what we are interested in. So the aim of this study is to find why customers repurchase a brand after unsatisfied experience and the research questions of this paper are:

(1) What factors can impact customers repurchase frequency of the same brand after unsatisfied experience?
(2) Which kind of brands do customers prefer to repurchase after unsatisfied experience?
(3) Why do customers repurchase these brands after unsatisfied experience?

We have obtained a large number of consumer repurchase records and reviews on Amazon's public database, and through data mining and data analysis, the existence of this phenomenon is preliminarily proved. In order to reduce the bias caused by omitted variables, we use fixed-effects models and control some related variables.

The structure of this paper is as follows: next chapter is the review of the literature on customer repurchase behavior. Then, we develop a framework of the determinants of online repurchase behavior after unsatisfied experience in online review contexts. Lastly we calculate and discuss the data analysis results.

2 Literature Review

2.1 Transaction-Specific Satisfaction

Customer satisfaction and related constructs are important to increase competition and marketing for a firm [8]. There are some theories and practice of customer satisfaction measurement [8]. According to Bitner and Hubbert [9], transaction-specific satisfaction refers to "the consumer's dis/satisfaction with a discrete service encounter." That is, consumers tend to comment on particular events of a transaction [10]. Some researchers make empirical investigation of transaction-specific satisfaction and overall satisfaction [9, 11, 12]. Jones and Suh [10] empirically investigate transaction-specific satisfaction, overall satisfaction and repurchase intentions and find that the two types of satisfaction can be distinguished from one another. Also, they find both types of satisfaction can be measured using the same scale, and transaction-specific satisfaction has an influence on repurchase intentions when overall satisfaction is controlled. Furthermore, Jones and Suh [10] show that the interaction between transaction-specific satisfaction and overall satisfaction has a negative effect on consumers' repurchase intentions, while in this model, the main effect on repurchase intentions of overall satisfaction is positive. Thus, we propose that:

H1: Transaction-specific satisfaction has a negative influence on customers' repurchase behavior after unsatisfied experience.

2.2 Overall Satisfaction

Mittal, Ross and Baldasare [13] find that overall satisfaction and attribute-level performance have separate and distinct effects on repurchase intentions. Previous research [14] shows that overall satisfaction and performance are related to repurchase behavior. Montoya-Weiss, Voss and Grewal [15] propose a conceptual model of the determinants of online channel use and overall satisfaction with the service provider. They find the service quality provided through both the online channel and the traditional channel affects customers' overall satisfaction. Some researchers [1, 16] find that higher levels of customer satisfaction rates result in higher levels of customer retention rates, which leads to the increase of customer repurchase behavior.

Even though customers have unsatisfied experiences, they will also repurchase if the service recovery is given to increase their overall satisfaction [3, 17], or if their families and friends persuade and convince them. Thus, repurchase behavior can exist after unsatisfied experience.

According to Bitner and Hubbert [9], overall satisfaction refers to "the consumer's overall dis/satisfaction with the organization based on all encounters and experiences with that particular organization". Jones and Suh [10] think customers view overall satisfaction as commenting on global impressions and general experiences with the firm. Thus, we propose that:

> H2: Overall satisfaction has a positive influence on repurchase behavior after unsatisfied experience.

2.3 Purchase Experience

Pappas et al. [18] describe that experience is indispensable factor for successful customer retention. Liang and Huang [19] point out that high-experienced customers tend to continue shopping. Zhou et al. [20] find that experience affects positively customers' intention to purchase online. However, Dholakia and Zhao [21] show that experienced customers can be hardly satisfied since they obtain more information during the shopping process.

According to Shannon's information theory [22], each thing has two states. When it is in a certain state S1 for a long time, then the probability $P(S_1) = p_1$ of the next moment at S_1 is large (may be set 0.9), and the probability $P(S_2) = p_2$ in state S_2 is small ($p_2 = 0.1$). The amount of information that event S_1 brings to people is $I_1 = -\log (p_1) = 0.152$ (takes 2 as the base), and the amount of information of S_2 is $I_2 = -\log (p_2) = 3.322$. Obviously I_1 is much smaller than I_2. That is to say, when the next moment of this thing is still at S_1, it brings little information to customers, so its influence on them is very small. On the contrary, when it is at S_2, because of the huge amount of information, its influence is very large, so everyone will pay attention to this matter. Thus, before they feel unsatisfied with a brand, customers who keep purchasing this brand are in satisfied emotion all the time. But when they suffer unsatisfied experiences for the first time, they may be impressed dramatically by this experience according to Shannon's information theory. Therefore, these customers' repurchase behavior of the same brand may decrease. Thus, we propose that:

H3a: Customers' prior online purchase experience has an influence on their repurchase behavior after unsatisfied experience.

H3b: Customers' satisfied purchase experience before they suffer unsatisfied experiences has a negative influence on repurchase behavior after unsatisfied experience.

2.4 Brand

Another key determinant of repurchase intent is public brand image [23]. Public brand image has been proved to affect repurchase intent directly [23] and indirectly through customer satisfaction [24–26]. Also, some researchers find that the positive effect of public brand image on repurchase intent is stronger for women than for men [27]. Brand preference is an intervening factor between customer satisfaction and repurchase intention [28]. Thus, brand has an influence on repurchase behavior. Meanwhile, Hellier et al. [28] suggest a model and show that there are seven important factors influencing repurchase intention, namely, service quality, equity and value, customer satisfaction, past loyalty, expected switching cost and brand preference. In addition, brand preference mediates the relationship between customer satisfaction and repurchase intention [28].

H4: Brand has a significant influence on customer repurchase behavior after unsatisfied experience.

3 Methodology

3.1 Fixed-Effects Model

A fixed-effects model is a statistical model and its parameters are fixed or non-random quantities, different from random effects models and mixed models. Regression methods of fixed effects are used to analyze longitudinal data with repeated measures on both independent and dependent variables [29]. Fixed-effects model makes it possible to control variables that have not or cannot be measured. For nonexperimental data, how to statically control variables that cannot be observed is difficult, unlike experimental research. Generally, three regression methods can construct fixed-effects models, namely, Least Square Dummy Variable model (LSDV), first-differenced equation (FD) and covariance estimator (CE).

Allison [30] points out that the dummy variable approach works well for linear regression and Poisson regression. It is acknowledged that the distribution of consumer repurchase behavior is Poisson distribution. And the fixed-effects analysis of repeated event data is conveniently [30]. Thus, this paper chooses LSDV method to construct our models. The benefit of LSDV is that an estimate of individual heterogeneity can be obtained to truly detect whether the fixed-effects model is effective.

3.2 Least Square Dummy Variable Model

Fixed-effects methods are used for data in which the dependent variable is measured on an interval scale and is linearly dependent on a set of predictor variables [29]. There are a set of individuals $(i = 1, \ldots, N)$, each of whom is measured at two or more points in time $(t = 1, \ldots, T)$. The basic model is:

$$y_{it} = \alpha + \beta X_{it} + \gamma Z_{it} + \varepsilon_{it} \tag{1}$$

Where y_{it} is the dependent variable, α is a constant, β and γ are vectors of coefficients, ε_{it} is for each individual at each point in time, X_{it} are predictor variables, and Z_{it} are dummy variables (Table 1).

Table 1. Symbol

Symbol	Description	Type
tss	Transaction-specific satisfaction	Numerical
pe	Purchase experience	Dummy
frp	Frequency of repurchase behavior of the same brand after unsatisfied experience	Numerical
ftpb	Frequency of total purchase behavior of the same brand	Numerical
fue	Frequency of unsatisfied experience	Numerical
osat	Overall satisfaction	Numerical
osv	Overall satisfaction variance	Numerical
suev	Satisfaction after unsatisfied experience variance	Numerical
$Brand_{it}$	Brand dummy variable	Dummy

Thus, we suggest our model:
Model A:

$$frp_1 = \alpha_1 + \beta_1 ftpb + \beta_2 fue + \beta_3 tss + \beta_4 osat + \beta_5 pe + \beta_6 osv + \gamma_1 Brand_{it} + \varepsilon_{it} \tag{2}$$

Model B:

$$\begin{aligned} frp_2 = {} & \alpha_2 + \beta_7 ftpb + \beta_8 fue + \beta_9 tss + \beta_{10} osat + \beta_{11} pe + \beta_{12} osv \\ & + \beta_{13} suev + \gamma_2 Brand_{it} + \varepsilon_{it} \end{aligned} \tag{3}$$

3.3   Data Collection

We use online Amazon review data [31, 32] to test the research models. Based on the research of Kincade et al. [33], the relationship between product durability and the repurchase of the brand variable is not significant, so this study chooses beauty industry as the object and the time span of the data is from 2003 to 2014.

There are 241,974 reviews in this sample, in which 869 reviews have missing values, so we exclude flawed reviews. The reviews we used are given by 182,624 customers, so repurchase behavior exists among these customers. Statistically, 81,062 customers, nearly half of the total, give negative reviews and among these customers, 14,901 have repurchase behavior.

To study the effect of brand on repurchase behavior after unsatisfied experience, the sample consists of the reviews of top20 brands.

Furthermore, to find out the real reasons why customers complain when they suffer unsatisfied experiences, we use 774,255 reviews, of which the customer rating are below 5 stars (5 stars are the highest score, and 1 star is the lowest), which we define as bad reviews, to solve this problem through text mining.

4   Results and Discussions

4.1   Models and Analysis

In order to test our hypotheses, whether transaction-specific satisfaction (H1), overall satisfaction (H2), purchase experience (H3) and brand (H4) influence customer repurchase behavior after unsatisfied experience, we use Least Square Dummy Variable model (LSDV) to construct fixed-effects (FE) estimation. Two models are used in this paper, the basic one (Model A) and the extended one (Model B). These two models regard frequency of repurchase behavior after unsatisfied experience as the dependent variable, and transaction-specific satisfaction, overall satisfaction, purchase experience and brand dummy variables as the independent variables. Frequency of total purchase behavior, frequency of unsatisfied experience and overall satisfaction variance are control variables. These variables above are the same in the two models, but in model B, there is another control variable: satisfaction after unsatisfied purchase experience variance, which can demonstrates the influence of satisfaction in different time on repurchase behavior after unsatisfied experience.

Table 2. Models

Independent variable	Dependent variable		Hypothesis
	Repurchase behavior		
	Model 1	Model 2	
Constant	−1.3294***	−1.3253***	
	(0.0290)	(0.0290)	
Control variable			
ftpb	0.8353***	0.8351***	
	(0.0024)	(0.0024)	
fue	0.2328***	0.2327***	
	(0.0046)	(0.0046)	
osv	0.0499***	0.0438***	
	(0.0053)	(0.0056)	
suev		0.0216**	
		(0.0066)	

(*continued*)

Table 2. (*continued*)

Independent variable	Dependent variable		Hypothesis
Explanatory variable			
tss	−0.0461***	−0.0438***	H1
	(0.0067)	(0.0067)	
osat	0.1418***	0.1396***	H2
	(0.0079)	(0.0080)	
pe	−0.2630***	−0.2637***	H3a
	(0.0025)	(0.0025)	H3b
F	20750.67	19966.23	
P	0.0000	0.0000	
R^2	0.9721	0.9721	
Root MSE	0.5235	0.5233	

$p < 0.001$*** $p < 0.005$** $p < 0.01$*

Table 2 presents the results of our fixed-effects analysis of the basic model (Model A) and of the extended model (Model B). The P of these two models are significant at 0.001 level, that is, these two models are both effective. The R^2 of these two models reaches 0.9721, which means that there are basically no omitted variables in the models. Both *Root MSE* are nearly 0.52 that is a smaller one. The smaller the value of *Root MSE* is, the better the regression effect will be. To sum up, brand effect is related to customer repurchase behavior after unsatisfied experience, which can confirm H4 through these two models.

Model A shows the basic model, covering several explanatory variables affecting customer repurchase behavior after unsatisfied experience. The findings of this model on the repurchase behavior variables are generally consistent with those of prior studies. We also find that transaction-specific satisfaction has a negative influence on customer repurchase behavior after unsatisfied experience, and that purchase experience can influence repurchase behavior after unsatisfied experience.

In model A, purchase experience is a significant predictor of customer repurchase behavior after unsatisfied experience. Its coefficient is −0.2630 ($P < 0.001$), negatively influencing the dependent variable. Meanwhile, overall satisfaction coefficient is 0.1418 ($P < 0.001$), positively affecting repurchase behavior after unsatisfied experience. And transaction-specific satisfaction negatively influences customer repurchase behavior after unsatisfied experience, and its coefficient is −0.0461 ($P < 0.001$). The opposite effects of these two factors reflect that different kind of satisfaction has different influence on repurchase behavior after unsatisfied experience. That is, customer repurchase behavior is indeed influenced by unsatisfied experience. Besides, overall satisfaction variance positively affects repurchase behavior after unsatisfied experience. Though the coefficient (0.0499) is small, its P is significant ($P < 0.001$).

Model B is developed based on Model A, which also reveals that brand has an influence on customer repurchase behavior after unsatisfied experience. It contains the same independent variables as Model A, and additionally includes another control factor, satisfaction after unsatisfied experience variance, which is related to different states of satisfaction. Repurchase behavior after unsatisfied experience is strongly influenced by purchase experience ($C. = -0.2637$, $P < 0.001$), followed by overall satisfaction ($C. = 0.1396$, $P < 0.001$), transaction-specific satisfaction ($C. = -0.0438$, $P < 0.001$), overall satisfaction variance ($C. = 0.0438$, $P < 0.001$) and satisfaction after unsatisfied experience variance ($C. = 0.0216$, $P < 0.001$), which can prove the accuracy and validity of Model A as well. Comparatively, satisfaction after unsatisfied experience variance has the least impact on repurchase behavior after unsatisfied experience.

In addition, the coefficients of control variables show that total purchase behavior and frequency of unsatisfied experience indeed positively affect customer repurchase behavior after unsatisfied experience. When customers suffer unsatisfied experiences, especially for the first time, their unsatisfied emotion has a strong effect on customers' intent and behavior. That is, unsatisfied emotion will change customers' attitude towards products next time when they shop.

Table 3. Model a brand effects significance

Brand (top 1–20)	Total sales	LSDV coefficient	Standard error	P
L'Oreal Paris	26269	−0.1977	0.0267	0.000
Conair	22258	−0.0853	0.0319	0.007
OPI	19604	−0.1801	0.0248	0.000
Olay	18320	−0.1232	0.0247	0.000
Revlon	17648	−0.0949	0.0256	0.000
Neutrogena	16852	−0.0798	0.0237	0.001
Maybelline	15421	−0.0658	0.0221	0.003
NYX	10641	−0.0507	0.0228	0.026
SHANY Cosmetics	10511	−0.0579	0.0264	0.028
Remington	8677	−0.0770	0.0272	0.005
HSI PROFESSIONAL	8651	−0.0629	0.0407	0.122
BaBylissPRO	7950	−0.0631	0.0329	0.048
Bare Escentuals	7924	−0.0578	0.0290	0.055
COVERGIRL	7814	−0.0464	0.0235	0.046
WEN® by Chaz Dean	7683	−0.0540	0.0277	0.049
Dove	7319	−0.0801	0.0244	0.051
Essie	7199	−0.0529	0.0260	0.001
Paul Mitchell	6866	−0.0582	0.0329	0.041
e.l.f. Cosmetics	6798	−0.0054	0.0224	0.810
Garnier[a]	6717			

[a]The last brand Garnier is adopted as the base brand.

As can be seen from Table 3, most of the brand dummy variables are significant ($P < 0.05$), so we can reject the null hypothesis that all brand dummy variables are 0. It indicates that brand has an effect on customer repurchase behavior after unsatisfied experience, and that mixed regression should not be used in this model.

Furthermore, the brand effect coefficients are all negative, which means the sales volume of a brand is inversely proportional to the frequency of repurchase behavior after unsatisfied experience. The table shows that brands with large sales volume have a more negative influence on repurchase behavior after unsatisfied experience than those with small sales, because the former's absolute value of the coefficients are larger. We find that after unsatisfied experience, the frequency of repurchase behavior on popular brand is less than that on unpopular brand. That is, customers prefer to repurchase unpopular brand after unsatisfied experience.

4.2   Robust Test

To test our model and to test whether the brand effect exists, we conduct Robust test on the research results.

First, we choose part of sample data to construct a fixed-effects model. The reason to do this is that if the fixed effect does exist, it should be independent of data. That is, whether a model is effective or not doesn't depend on its amount of data. Then, we use these data to construct a random-effects model.

There are 9972 reviews in this sample, and we first carry out the Hausman test. The results are presented as follows:

Table 4. Hausman test results

	Fixed-effects	Random-effects	Difference	S.E.
Constant	−1.4321	−1.4524	0.0203	0.0042
Control variable				
ftpb	0.8099	0.8117	−0.0018	0.0003
fue	0.2589	0.2617	−0.0028	0.0006
osv	0.0414	0.0486	−0.0072	0.0009
suev	0.0259	0.0106	0.0153	0.0019
Explanatory variable				
tss	−0.0570	−0.0595	0.0025	0.0006
pe	−0.2553	−0.2603	0.0050	0.0006
osat	0.1661	0.1713	−0.0052	0.0010
chi(2)	85.11			
P	0.0000			

From Table 4, since the P is 0.000, the result is strongly significant, so a fixed-effect model, rather than a random-effects model, should be used.

From Tables 3 and 4, we can see that most of dummy variables are significant, but the rest are not. So we use joint significance test and examine the joint significance of all the dummy variables. The result F is 6.99, and P is 0.000, which confirms that brand effects should be included in the model.

Additionally, considering that time may have an impact on the model, to avoid omitting variable bias, we also use the total data to construct a time fixed-effects model,

Table 5. Time fixed-effects significance

Time	P
2014	0.251
2013	0.262
2012	0.250
2011	0.257
2010	0.241
2009	0.309
2008	0.287
2007	0.303
2006	0.395
2005	0.508
2004	0.721

*2003 is adopted as the base period.

considering that time may have an impact on the model. But from Table 5, the results show that all time dummy variables are not significant because all P values are large. So we can conclude that there is no time effect.

4.3   Text Mining

To find out what factors influence customers to repurchase a brand after unsatisfied experience, we use text mining to find out the key which customers truly care about. Based on beauty industry reviews, we use term-frequency analysis through Python 3.0. There are 134,531 words given by customers in online reviews.

We exclude several words like "the", "was", "in" and so on.

Table 6. High-frequency words (top 1–40)

Words 1–10	Times	Words 11–20	Times	Words 21–30	Times	Words 31–40	Times
Hair	416142	Price	75023	Shampoo	53070	Conditioner	37173
Product	397379	Works	71050	Smells	47929	Quality	36520
Like	302478	Bottle	67839	Cream	45213	Soft	36165
Skin	191057	Look	66484	Light	44549	Thick	35996
Color	106645	Try	63487	Oil	44129	Disappointed	35832
Smell	97360	See	60517	Amazon	41194	Hard	35216
Love	83817	Feel	60438	Small	41094	Natural	34538
Face	82648	Brush	58055	Lotion	40870	Worth	33744
Dry	77600	Scent	56168	Purchased	38973	Old	33512
Products	75802	Money	54449	Reviews	38643	Wash	32432

From Table 6, these words are related to the reasons why customers repurchase a brand after unsatisfied experience. Most customers may lay more emphases on whether products are useful to their hair, skin or face, whether the color of products looks well, whether the price is suitable and whether exterior design matches product. These factors are more related to product itself when customers repurchase the same brand after they suffer unsatisfied experiences. To sum up, we find that product quality, product efficacy, product packaging, price, reviews and purchase experience can influence customers' repurchase behavior after unsatisfied experience.

In addition, the word "like" is used about 302,478 times and the word "love" is used 83,817 times, which means in bad reviews, customers do not always complain about products with negative words, but positive words will also be given. These words are useful to managers and for firms.

5   Conclusion

The paper proposes and tests fixed-effects models through LSDV method to examine the factors influencing customer repurchase behavior after unsatisfied experience. In doing so, it provides a theoretical and empirical improvement for prior studies on repurchase behavior. It additionally provides another new factor due to its particularity; for instant, influence factor transaction-specific satisfaction will appear after unsatisfied experience.

Most importantly, this study finds that the brand has an important impact on the repurchase behavior after unsatisfied experience, and that there exists brand effect in repurchase behavior after unsatisfied experience. Through the LSDV method, the sales volume of the brand is related to the repurchase behavior after unsatisfied experience. The higher the brand sales volume is, the less frequent repurchase behavior after unsatisfied experience will be. This may be related to the brand attributes [34].

In addition, from the perspective of control variables, the coefficients of them are relatively large. The reason is possible that the total number of purchase can reflect the purchase power of consumers, the customers' income. Because Seiders and Voss [35] find that income and repurchase behavior are positively correlated.

Finally, this study finds the factors influencing customers repurchase behavior of the same brand after unsatisfied experience are more related to product itself, but less to channel.

6   Limitations and Future Directions

According to the models, it is clear that brand has a strong influence on repurchase behavior after unsatisfied experience. However, due to data limitations, we only suggest a small number of factors about how brand affects customer repurchase behavior after unsatisfied experience. It is necessary to conduct an in-depth study on how brand affects customer repurchase behavior after unsatisfied experience, namely, what its internal mechanism is and how it works.

This study does not systematically study the internal psychological mechanism affecting consumer repurchase behavior. According to the previous research, different generations of customers react differently to the same thing [3]. After the service failure, customers' complaints and repurchase behaviors are also different.

References

1. Reichheld, F.F., Sasser, J.W.: Zero defections: quality comes to services. Harv. Bus. Rev. **68**(5), 105–111 (1990)
2. Knox, G., Van Oest, R.: Customer complaints and recovery effectiveness: a customer base approach. J. Mark. **78**(5), 42–57 (2014)
3. Soares, R.R., Zhang, T.T., Proença, J.F., Kandampully, J.: Why are Generation Y consumers the most likely to complain and repurchase? J. Serv. Manag. **28**(3), 520–540 (2017)
4. Wu, L.Y., Chen, K.Y., Chen, P.Y., Cheng, S.L.: Perceived value, transaction cost, and repurchase-intention in online shopping: a relational exchange perspective. J. Bus. Res. **67**(1), 2768–2776 (2014)
5. Wen, C., Prybutok, V.R., Xu, C.: An integrated model for customer online repurchase intention. J. Comput. Inf. Syst. **52**(1), 14–23 (2011)
6. Cudd, M., Davis, H.E., Eduardo, M.: Mimicking behavior in repurchase decisions. J. Behav. Financ. **7**(4), 222–229 (2006)
7. Mao, Z., Lyu, J.: Why travelers use Airbnb again? An integrative approach to understanding travelers' repurchase intention. Int. J. Contemp. Hosp. Manag. **29**(9), 2464–2482 (2017)
8. Babin, B.J., Griffin, M.: The nature of satisfaction: an updated examination and analysis. J. Bus. Res. **41**(2), 127–136 (1998)
9. Bitner, M.J., Hubbert, A.R.: Encounter satisfaction versus overall satisfaction versus quality. Serv. Qual. New Dir. Theory Pract. **34**(2), 72–94 (1994)
10. Jones, M.A., Suh, J.: Transaction-specific satisfaction and overall satisfaction: an empirical analysis. J. Serv. Mark. **14**(2), 147–159 (2000)
11. Anderson, E.W., Fornell, C.: A customer satisfaction research prospectus. Serv. Qual. New Dir. Theory Pract. **14**(1), 239–266 (1994)
12. Parasuraman, A., Zeithaml, V.A., Berry, L.L.: Reassessment of expectations as a comparison standard in measuring service quality: implications for further research. J. Mark. **58**(1), 111–124 (1994)
13. Mittal, V., Ross Jr., W.T., Baldasare, P.M.: The asymmetric impact of negative and positive attribute-level performance on overall satisfaction and repurchase intentions. J. Mark. **62**(1), 33–47 (1998)
14. Oliva, T.A., Oliver, R.L., MacMillan, I.C.: A catastrophe model for developing service satisfaction strategies. J. Mark. **56**(3), 83–95 (1992)
15. Montoya-Weiss, M.M., Voss, G.B., Grewal, D.: Determinants of online channel use and overall satisfaction with a relational, multichannel service provider. J. Acad. Mark. Sci. **31**(4), 448–458 (2003)
16. Hogan, J.E., Lemon, K.N., Libai, B.: What is the true value of a lost customer? J. Serv. Res. **5**(3), 196–208 (2003)
17. Rust, R.T., Zahorik, A.J.: Customer satisfaction, customer retention, and market share. J. Retail. **69**(2), 193–215 (1993)
18. Pappas, I.O., Pateli, A.G., Giannakos, M.N., Chrissikopoulos, V.: Moderating effects of online shopping experience on customer satisfaction and repurchase intentions. Int. J. Retail. Distrib. Manag. **42**(3), 187–204 (2014)

19. Liang, T.P., Huang, J.S.: An empirical study on consumer acceptance of products in electronic markets: a transaction cost model. Decis. Support Syst. **24**(1), 29–43 (1998)
20. Zhou, L., Dai, L., Zhang, D.: Online shopping acceptance model - a critical survey of consumer factors in online shopping. J. Electron. Commer. Res. **8**, 41–61 (2007)
21. Roy Dholakia, R., Zhao, M.: Effects of online store attributes on customer satisfaction and repurchase intentions. Int. J. Retail. Distrib. Manag. **38**(7), 482–496 (2010)
22. Shannon, C.E.: A mathematical theory of communication. ACM SIGMOBILE Mob. Comput. Commun. Rev. **5**(1), 3–55 (2001)
23. Johnson, M.D., Gustafsson, A., Andreassen, T.W., Lervik, L., Cha, J.: The evolution and future of national customer satisfaction index models. J. Econ. Psychol. **22**(2), 217–245 (2001)
24. Ball, D., Simões Coelho, P., Machás, A.: The role of communication and trust in explaining customer loyalty: an extension to the ECSI model. Eur. J. Mark. **38**(9/10), 1272–1293 (2004)
25. Ball, D., Coelho, P.S., Vilares, M.J.: Service personalization and loyalty. J. Serv. Mark. **20**(6), 391–403 (2006)
26. Türkyılmaz, A., Özkan, C.: Development of a customer satisfaction index model: an application to the Turkish mobile phone sector. Ind. Manag. Data Syst. **107**(5), 672–687 (2007)
27. Frank, B., Enkawa, T., Schvaneveldt, S.J.: How do the success factors driving repurchase intent differ between male and female customers? J. Acad. Mark. Sci. **42**(2), 171–185 (2014)
28. Hellier, P.K., Geursen, G.M., Carr, R.A., Rickard, J.A.: Customer repurchase intention: a general structural equation model. Eur. J. Mark. **37**(11/12), 1762–1800 (2003)
29. Allison, P.: Fixed Effects Regression Methods in SAS. SAGE, Thousand Oaks (2009)
30. Allison, P.: Bias in fixed-effects Cox regression with dummy variables. Manuscript, Department of Sociology, University of Pennsylvania (2002)
31. He, R., McAuley, J.: Ups and downs: modeling the visual evolution of fashion trends with one-class collaborative filtering. In: 25th International Conference Proceedings on World Wide Web, pp. 507–517. International World Wide Web Conferences Steering Committee, Switzerland (2016)
32. McAuley, J., Targett, C., Shi, Q., Van Den Hengel, A.: Image-based recommendations on styles and substitutes. In: 38th International ACM SIGIR Conference Proceedings on Research and Development in Information Retrieval, pp. 43–52. ACM, New York (2015)
33. Kincade, D.H., Giddings, V.L., Chen-Yu, H.J.: Impact of product-specific variables on consumers' post-consumption behaviour for apparel products: USA. J. Consum. Stud. Home Econ. **22**(2), 81–90 (1998)
34. Erciş, A., Ünal, S., Candan, F.B., Yıldırım, H.: The effect of brand satisfaction, trust and brand commitment on loyalty and repurchase intentions. Procedia-Soc. Behav. Sci. **58**, 1395–1404 (2012)
35. Seiders, K., Voss, G.B., Grewal, D., Godfrey, A.L.: Do satisfied customers buy more? Examining moderating influences in a retailing context. J. Mark. **69**(4), 26–43 (2005)

How to Sell Your House for More?

Guohou Shan[1(✉)], Dongsong Zhang[2], Lina Zhou[2], and James Clavin[1]

[1] University of Maryland Baltimore County, Baltimore, MD 21250, USA
{gshan2, jclavin}@umbc.edu
[2] University of North Carolina at Charlotte, Charlotte, NC 28223, USA
{dzhang15, lzhou8}@uncc.edu

Abstract. Online auction is one of the common mechanisms for online selling and buying. Despite a host of studies on influencing factors on online auction, there is an insufficient understanding of the effects of item features and auction characteristics on the final price. In this study, we extracted both verbal and nonverbal features of online house auctions and examined their effects on the final price of auctioned houses. Based on an analysis of 10,573 online house auctions, we found that features such as starting price have a positive impact on the final price, and features such as bid increment have a negative effect. In addition, verbal features such as numeral count also negatively influence the final auction price. The findings have broad practical implications for improving the description and design of online auction items.

Keywords: Online house auction · Feature importance ·
Natural language processing · Regression analysis

1 Introduction

Online auctions are becoming increasingly popular by attracting more and more sellers and buyers. Many studies have investigated underlying economic mechanisms, trust or reputation models, and explanation of the final price of online auctions to date [13, 22]. However, the effects of product features on the final price in online auction remain unclear. Understanding the issue can facilitate both sellers in improving the design of online auctions and item descriptions and bidders in offering a fair price, which would further contribute to the success of online auction market.

In view of different types of auction items, we focus on house auction in this study. We selected house auction for two main reasons: (1) houses are considered as expensive items relative to other typical auctioned items, which heightens the need of risk management; and (2) houses have been relatively less studied compared with other types of auction items in the literature. Our overarching research question is what factors contribute to house price in online auction.

To address the research question, we examine the effects of online house auction features related to house description and auction on the final auction price. These factors are explored from both verbal (e.g., features extracted from text descriptions) and nonverbal (e.g., starting price, bidding increment, appraisal price) perspectives to gain a fuller understanding of their effects. In addition, we also assess the importance of

© Springer Nature Switzerland AG 2019
J. J. Xu et al. (Eds.): WEB 2018, LNBIP 357, pp. 91–98, 2019.
https://doi.org/10.1007/978-3-030-22784-5_9

individual features for explaining the final house price. To the best of our knowledge, this is the first study that investigates the impact of individual verbal and nonverbal features of online auction on the final price of auctioned houses. Further, we conducted empirical evaluations of the extracted features using real world data collection from one of the largest online auction platforms in China, Ali Auction1. The findings of this empirical study can help improve the state of knowledge on contributing factors to online auction price and facilitate various stakeholders of online auction in making better decisions.

The rest of this paper is organized as follows. First, we introduce related work on online auction trust models and price prediction. Then, we present the proposed methods, followed by the reporting of data analysis and results. Next, we discuss research findings, contributions, and practical implications of this study.

2 Related Studies

Online auctions have been investigated extensively in the past couple of decades. Based on research objectives, we classify existing research into three broad categories, including trust models for online auction, online auction characteristics, and online auction price prediction.

Research on developing trust models for online auctions mainly focuses on trust/reputation system design, bidding justice and trust effects on selling, and influential factors on trust [2, 3, 5, 9, 14, 18, 20–22]. For example, Lin et al. [14] proposed a reputation system in building trust models in online auctions that considered the reputations of sellers and buyers, which outperformed several existing reputation systems in selecting honest sellers in simulation experiments. Tu et al. [18] studied selling strategies of online auction by segmenting the market from the perspectives of trust and information asymmetry. Accordingly, they classified auction listings into four segments: new vs. experienced sellers (i.e., trust), and new vs. used items (i.e., information asymmetry) and found that trust was more important to auction success than transaction enhancing strategies. Wu et al. [20] investigated the factors that influenced bidders' trust. They conducted laboratory experiments by manipulating the trust arguments of "about me" pages and administering online surveys. The results showed that the trust of bidders was significantly influenced by benevolent sellers, structural assurance, and disposition to trust.

Research on online auction characteristics has mainly focused on price comparisons and sellers' listing strategy [3, 4, 6–8]. For instance, Chow et al. [6] compared auctions and negotiated sales in theoretical and empirical aspects. They built a theoretical model with the data collected from property sales in Singapore and found that auctions resulted in higher prices than negotiated sales when demand for the asset was strong, and that the auction mechanism would obtain higher prices in an "up market" than in a flat market [6]. Chen et al. [4] investigated sellers' optimal listing strategies in online auctions and proposed a unified framework to characterize them. Through a mathematical analysis, they found that different types of sellers would likely adopt different auction strategies. For example, sellers under a medium range of time pressure adopted the buy-it-now auction strategy, while the most impatient sellers adopted the fixed-price listing.

Research on online auction price prediction mainly employ various machine learning algorithms and non-machine learning models [10–13, 19, 23]. For example, Khadge et al. [11] developed a machine learning based online auction price predictor using eBay online auction data and Naïve Bayes and Support Vector Machine (SVM). Their results showed that Naïve Bayes achieved a 99.33% accuracy in predicting online auction success, and that SVM achieved a 96.3% accuracy as to whether a product maximized its profit or not. Kaur et al. [10] designed an automated dynamic bidding agent using fuzzy reasoning techniques to assist bidders in making decisions per their bidding behaviors. By using an empirical evaluation, they demonstrated that the bidding agent lead to improved initial price and the proficiency of the fuzzy bidding strategies in terms of the success rate and expected utility. Kuruzovich and Etzion [13] investigated sellers' pricing decisions and auction outcomes in the context of multi-channel retailing. They developed an analytical model using search theory and utilized data collected from eBay Motors. The study revealed that an increase of bidders' valuations and a seller's auction reserve price could lead to a positive association between seller characteristics and auction price.

Despite exiting research on online auction, as described above, there is a lack of research studying the effects of features of auction and auction item description on the final auction price, especially from both verbal and non-verbal perspectives, which can be critical to helping sellers improve their online auction performance, to helping bidders formulate better bidding strategies, or even to laying the groundwork for building predictive models for final auction price.

3   Method

This study followed the research process as depicted in Fig. 1.

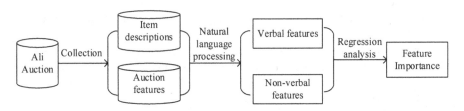

Fig. 1. Research method.

- First, we crawled transaction data from Ali Auction, including house description and auction settings, etc., for 18 months starting from January 2017. There were a total of 10,573 completed transactions. A sample Ali house auction page is shown in Fig. 2.
- Then, we performed natural language processing to extract both verbal and non-verbal features from the crawled data. This research mainly utilized the Part-of-Speech (POS) tagging and sentiment analysis to extract verbal and non-verbal features from online house auctions and house descriptions, including verbal

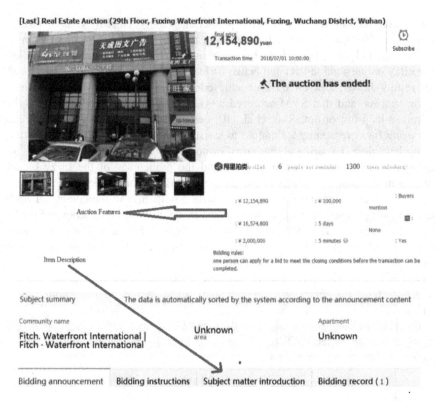

Fig. 2. A sample of house auction at ali house auction (translated version).

features such as house description, sentiment polarity, and the number of aggregated adjectives used in the house description. POS tagging is a process of determining the property of words in text [1]. Sentiment analysis is a process of extracting sentiments from text [15], which has been commonly applied to online content to discover people's attitude, emotions, and feelings toward an entity [17].

- Finally, we constructed a multiple linear regression model to predict final auction price using the extracted features and examined the importance of individual features for explaining the final auction price.

4 Data Analysis and Results

Given the fact that very scarce existing studies has investigated the effects of online auction features on the final auction price, we extracted a large number of features that potentially have such an explanatory power. It also allowed us to explore which features have stronger effects than others. At the end, we extracted a total of 47 features (See Table 1). These features were grouped into verbal and nonverbal categories. Among them, f0 and f12–f44 belong to verbal features, while the remaining ones are non-verbal features.

Table 1. A summary of features.

Category	Features
Verbal	Length of title (f0), Length of description (HD) (f12), Number of Sentence in HD (f13), Number of words in HD (f14), Sentiment score of HD (f15), Ratio of Morphological morpheme (f16), Ratio of adjectives (f17), Sub-word ratio (f18), Distinguishing word ratio (f19), Conjunction word ratio (f20), Adverb word ratio (f21), Position word ratio (f22), Morpheme word ratio (f23), Connection component (f24), Phrase (f25), Abbreviation word ratio (f26), Sequential ingredients (f27), Idiom (f28), Numeral word ratio (f29), Noun morpheme word ratio (f30), Noun word ratio (f31), Onomatopoeia word ratio (f32), Preposition (f33), Quantitative word ratio (f34), Pronoun (f35), Locality Category word ratio (f36), Time morpheme word ratio (f37), Time word ratio (f38), Auxiliary word ratio (f39), Dynamic morpheme (f40), Verb word ratio (f41), Not morpheme word ratio (f42), Modal particle word ratio (f43), Statement word ratio (f44)
Nonverbal	Multiple houses (f1), Starting price (f2), Appraisal price (f3), Deposit (f4), Bid Increment (f5), Bidding period (f6), Number of sign-ups (f7), Number of reminding (f8), Number of onlookers (f9), Number of delays (f10), Number of auction rounds (f11), Video status (f45), Number of images (f46)

Regression analysis is the most widely used statistical tool for discovering relationships among variables [16]. In this research, we conducted multiple linear regression analysis to explain the final online house auction price with online house auction features. The results are reported in Table 2. In addition to t-statistic, we also reported feature coefficients (weights) and standard errors (Std. error) in the table.

Table 2. Weighted least squares regression estimation results.

V	Coefficients	Std. error	t-Stat
I	$-3.34E + 05$	$1.44E + 05$	-2.327*
f0	$3.07E + 03$	$1.76E + 03$	1.739
f1	$-5.43E + 04$	$4.61E + 04$	-1.18
f2	$1.16E + 00$	$1.11E\text{-}02$	104.527***
f3	$-1.62E\text{-}01$	$6.79E\text{-}03$	-23.772***
f4	$1.36E + 00$	$3.53E\text{-}02$	38.525***
f5	$-3.09E + 00$	$2.52E\text{-}01$	-12.252***
f6	$1.68E + 00$	$2.83E + 01$	0.059
f7	$5.50E + 03$	$4.28E + 03$	1.283
f8	$-1.10E + 03$	$2.10E + 02$	-5.209***
f9	$2.90E + 01$	$3.53E + 00$	8.217***
f10	$2.53E + 03$	$9.69E + 02$	2.616**
f11	$6.66E + 03$	$9.06E + 02$	7.35***
f12	$-7.57E + 01$	$1.85E + 02$	-0.409

(continued)

Table 2. (*continued*)

V	Coefficients	Std. error	t-Stat
f13	−4.24E + 02	2.65E + 03	−0.16
f14	2.52E + 02	3.70E + 02	0.683
f15	NA	NA	NA
f16	−2.92E + 07	2.32E + 07	−1.258
f17	NA	NA	NA
f18	NA	NA	NA
f19	−5.05E + 05	3.28E + 06	−0.154
f20	2.84E + 06	2.86E + 06	0.992
f21	−1.33E + 06	2.43E + 06	−0.547
f22	−3.31E + 06	3.20E + 06	−1.033
f23	−4.74E + 06	1.75E + 07	−0.271
f24	6.09E + 07	3.08E + 07	1.98*
f25	4.96E + 06	7.19E + 06	0.689
f26	9.74E + 05	2.30E + 06	0.423
f27	5.62E + 07	4.02E + 07	1.397
f28	−5.00E + 06	4.52E + 06	−1.104
f29	−1.29E + 06	5.36E + 05	−2.403*
f30	−2.21E + 06	7.70E + 06	−0.287
f31	1.64E + 05	4.19E + 05	0.39
f32	5.54E + 07	2.55E + 08	0.217
f33	1.63E + 06	1.49E + 06	1.096
f34	−1.29E + 06	2.06E + 06	−0.623
f35	1.04E + 06	2.23E + 06	0.466
f36	−3.07E + 06	4.15E + 06	−0.739
f37	1.72E + 07	1.18E + 07	1.461
f38	8.19E + 05	5.73E + 06	0.143
f39	−2.94E + 06	5.41E + 06	−0.544
f40	4.89E + 06	1.49E + 07	0.328
f41	2.29E + 05	5.38E + 05	0.425
f42	1.58E + 05	2.13E + 05	0.741
f43	1.66E + 07	5.37E + 07	0.309
f44	−4.38E + 06	2.28E + 07	−0.192
f45	−6.72E + 04	3.28E + 04	−2.049*
f46	−3.98E + 03	5.93E + 03	−0.672

The adjusted R square is 0.9724, which shows that our model can explain the final price of auctioned houses extremely well. Among other features, f2, f4, f9, and f11 ($p < 0.001$), f10 ($p < 0.01$), and f24 ($p < 0.05$) had a positive effect on the final auction price, and f3, f5, and f8 ($p < 0.001$), f29 and f45 ($p < 0.05$) had a negative effect on the final price. Other features did not show a significant impact.

5 Discussion

This study extracted 47 verbal and non-verbal features from the online house auction transactions. Using the extracted features, we built a multiple linear regression model to explain the final auction price and to analyze the strength of effect of individual features on the final house price. The results show that different verbal and nonverbal features have varying effects on the final auction. In particular, some nonverbal features were found to have strong positive (negative) effects such as starting price and deposit (e.g., appraisal price, bid increment) while the verbal features, connection component and numeral word ratio, were found to have positive and negative effects respectively.

This study makes a few novel research contributions and practical implications. First, it examines the effects of features of online auction and house description on the final price. The findings could help sellers better describe their auction items and/or improve online auction process, such as setting a reasonable starting price that positively affects the final auction price. Second, this research innovatively classifies features of an online auction into two categories: verbal and non-verbal. The results demonstrated that some verbal features, such as the connection component word ratio (f24) and numeral word ratio (f29), of house description can influence the final house auction price, implying that researchers and practitioners should pay attention to the text descriptions of auctioned items. Third, the findings of this research can be used to improve the performance of predictive models for online auction price. For instance, incorporating features such as the starting price can potentially improve the performance of price prediction models.

This study has some limitations that offer potential opportunities for future research. We used data collected from a single online auction site. The generalizability of the proposed model and of the identified influential features on the final auction price to other online auction sites requires further research efforts. We focused on house auction in this study. The extension of the model for the final price as validated in this study to other types of auction items are worthy of separate investigation in future.

Acknowledgements. This research is supported in part by the National Science Foundation under Grant No. SES-152768, CNS-1704800. Any opinions, findings, and conclusions or recommendations expressed here are those of the authors and do not necessarily reflect the views of the National Science Foundation.

References

1. Brill, E.: Part-of-speech tagging. In: Handbook of Natural Language Processing, pp. 403–414 (2000)
2. Carter, M., Tams, S., Grover, V.: When do i profit? uncovering boundary conditions on reputation effects in online auctions. Inf. Manag. **54**(2), 256–267 (2017)
3. Chang, H.H.: Intelligent agent's technology characteristics applied to online auctions' task: a combined model of TTF and TAM. Technovation **28**(9), 564–577 (2008)
4. Chen, K.-P., Ho, S.-H., Liu, C.-H., Wang, C.-M.: The optimal listing strategy in online auctions. Int. Econ. Rev. **58**, 421–437 (2017)

5. Chiu, C.-M., Huang, H.-Y., Yen, C.-H.: Antecedents of trust in online auctions. Electron. Commer. Res. Appl. **9**(2), 148–159 (2010)
6. Chow, Y.L., Hafalir, I.E., Yavas, A.: Auction versus negotiated sale: evidence from real estate sales. Real Estate Econ. **43**(2), 432–470 (2015)
7. Einav, L., Farronato, C., Levin, J., Sundaresan, N.: Auctions versus posted prices in online markets. J. Polit. Econ. **126**(1), 178–215 (2018)
8. Gregg, D.G., Walczak, S.: Dressing your online auction business for success: an experiment comparing two ebay businesses. Mis Q. **32**(3), 653–670 (2008)
9. Gregg, D.G., Walczak, S.: The relationship between website quality, trust and price premiums at online auctions. Electron. Commer. Res. **10**(1), 1–25 (2010)
10. Kaur, P., Goyal, M., Lu, J.: A price prediction model for online auctions using fuzzy reasoning techniques. In: 2014 IEEE International Conference on Fuzzy Systems (FUZZ-IEEE), pp. 1311–1318. IEEE (2014)
11. Khadge, M.R., Kulkarni, M.V.: Machine learning approach for predicting end price of online auction. In: International Conference on Inventive Computation Technologies (ICICT), vol. 3, pp. 1–5. IEEE (2016)
12. Kumar, S., Rishi, R.: Hybrid dynamic price prediction model in online auctions. Int. J. Appl. Eng. Res. **12**(5), 598–604 (2017)
13. Kuruzovich, J., Etzion, H.: Online auctions and multichannel retailing. Manag. Sci. **64**(6), 2734–2753 (2018)
14. Lin, I.-C., Wu, H.-J., Li, S.-F., Cheng, C.-Y.: A fair reputation system for use in online auctions. J. Bus. Res. **68**(4), 878–882 (2015)
15. Liu, B.: Sentiment analysis and opinion mining. Synth. Lect. Hum. Lang. Technol. **5**(1), 1–167 (2012)
16. Neter, J., Wasserman, W., Kutner, M.H.: Applied linear regression models (1989)
17. Shan, G., Zhang, D., Zhou, L., Suo, L., Lim, J., Shi, C.: Inconsistency investigation between online review content and ratings (2018)
18. Tu, Y., Alex Tung, Y., Goes, P.: Online auction segmentation and effective selling strategy: trust and information asymmetry perspectives. J. Electron. Commer. Res. **18**(3), 189–211 (2017)
19. Wellman, M.P., Sodomka, E., Greenwald, A.: Self-confirming price prediction strategies for simultaneous one-shot auctions. arXiv preprint arXiv:1210.4915 (2012)
20. Wu, C.-S., Cheng, F.-F., Yen, D.C.: The influence of seller, auctioneer, and bidder factors on trust in online auctions. J. Organ. Comput. Electron. Commer. **24**(1), 36–57 (2014)
21. Xu, H., Shatz, S.M., Bates, C.K.: A framework for agent-based trust management in online auctions. In: Fifth International Conference on Information Technology: New Generations, ITNG 2008, pp. 149–155. IEEE (2008)
22. Yu, W.: Analysis on trust influencing factors and trust model from multiple perspectives of online Auction. Open Phys. **15**(1), 613–619 (2017)
23. Zhang, S., Jank, W., Shmueli, G.: Real-time forecasting of online auctions via functional k-nearest neighbors. Int. J. Forecast. **26**(4), 666–683 (2010)

How Do Novice Consumers Learn from Online Expert Reviews?

Zhuolan Bao[1(✉)] and Michael Chau[2]

[1] School of Management and Economics, CUHK Business School,
The Chinese University of Hong Kong, Shenzhen, China
baozhuolan@cuhk.edu.cn
[2] Faculty of Business and Economics,
The University of Hong Kong, Pok Fu Lam, Hong Kong
mchau@business.hku.hk

Abstract. Expert-written product reviews are prevalent and could be in many forms, ranging from short textual descriptions to video-embedded blogs. For consumers, expert reviews can not only optimize their purchase decisions, but also play an important role in facilitating consumer learning. In this paper, we first draw on profession research and propose two focal elements that characterize the content of expert reviews, and then investigate the impact of expert reviews on consumers' product preferences and information consumption via the lens of preference construction theory and dissonance theory. With describing our experiment design to test the hypotheses, we seek to make contributions to the vast studies of online reviews by characterizing the expert review content, examining the effects of expert review on consumer judgments and offering a new perspective of distinguishing expert reviews and peer reviews in online markets.

Keywords: Expert reviews · Preference construction ·
Information consumption · Peer reviews · Online markets

1 Introduction

Purchase decision-making of a new product is a difficult and complex job for novice consumers, as they have little knowledge about the category and no idea of how to gather and organize product information [1]. According to preference construction theory, when attribute information is uncertain, consumers tend to build their attribute preferences by learning about product information that is most accessible to them [2]. Internet naturally nurtures such an environment that enables a consumer to obtain a large amount of product information. Product reviews are one of the most important information sources [3], and are extensively acquired and consumed by consumers to facilitate purchase decisions.

Among massive product reviews, expert reviews are of a special type, which contains reviews written by professionals and could be in many forms, such as textual posts [4], blogs [5], videos [6], or reviews with expert labels in online markets [7].

© Springer Nature Switzerland AG 2019
J. J. Xu et al. (Eds.): WEB 2018, LNBIP 357, pp. 99–107, 2019.
https://doi.org/10.1007/978-3-030-22784-5_10

Prior research findings have revealed that expert reviews are influential to product consumption [7, 8], as well as the product's perceived quality [5].

Therefore, acknowledging the importance of expert reviews, many companies have started to endorse or ask experts to write positive reviews for their products [9]. But the explanations of expert reviews' impact on consumer perceptions are largely from the perspective of information credibility cues, which is more suitable for low involvement situations [10], or single-attribute product evaluation (i.e. using general attitudes, summary impressions or heuristics of the product to make judgement [11, 12]), such as when evaluating movie or book products. When it comes to a multi-attribute product, the evaluation needs consumers' higher level of involvement, and consumers would spend a large amount of time and efforts to make the right decisions [13].

Intuitively thinking, expert reviews should be more important in high-involvement purchase situations, given its potential in assisting consumers to understand the product attributes and relevant information. One relevant study by Gu and Park [14] also concluded that when evaluating a high-involvement product, external word-of-mouth (e.g. expert reviews) are more influential than internal word-of-mouth (e.g. consumer reviews on retailer-hosted websites). However, current research are largely neglecting investigations to such findings.

In this work, we set out to study the nature of expert reviews, and the mechanism of their impact on novice consumers in multi-attribute product purchases. We ask the following questions: (1) What characterize the content of expert reviews? (2) How do consumers learn from the expert review content? (3) Do expert reviews influence the consumption of peer reviews? To answer the questions, we start from profession studies and research on consumer preference construction, and then present a framework to investigate the distinctive role of expert reviews on consumers' preference construction and subsequent information consumption.

The remainder of the paper is arranged as follow. First, we will review the literature of expert reviews, and propose the characteristics of expert review content based on profession research. Next, we will develop our hypotheses via the lens of the preference construction theory and dissonance theory. After descriptions of our proposed methodology, we will discuss our potential contribution, limitations and opportunities for future studies.

2   Theory Development and Hypothesis Testing

2.1   Expert Reviews

Both expert reviews and peer-consumer reviews provide product experiences and evaluation, and both are shown to be influential in product sales or purchase intention [8]. Existing studies often differentiate the two by the size of their impact on consumer judgment. For example, review's expertise information is regarded as a peripheral cue which affects the persuasion process from the perspective of elaboration likelihood model [15]. Comparing to central cue information such as argument quality, review expertise is more influential when the product is of low personal relevance [10]. But when a product invokes more consumers' cognitive efforts, the central cue information is more important for consumers' judgment. Besides the elaboration likelihood model, an alternative view emerges with the idea of perceived distance of information. It stated that people tend to take the words from others who are like themselves [16]. Therefore, the reviews or recommendations written by peers are perceived as more valuable than those written by professions [16, 17].

However, both explanations are catering to understanding people's reliance on reviews for low involvement product evaluation [11]. For multi-attribute products, a deeper inspection should be more appropriate to comprehend the role of reviews. We draw on studies of professions, and provide a third theoretical lens of analyzing the impact of expert reviews.

2.2   Profession in Reviews

Professionals are individuals who are specialized in certain knowledge area and willing to serve the public [18]. In the seminal work in educational research, Freidson proposed two broad features that characterize profession, (a) acquisition and especially trained application of an unusually complex body of knowledge and skills, and (b) an objective of serving the needs of the public, with particular emphasis on an ethical or altruistic approach toward clients [19]. Therefore, with a professon's review, we expect the review to deliver the two features, domain knowledge or skills of evaluating the product, and principled evaluation of the product.

Similarly, Alba and Hutchinson [20] defined consumer expertise in a broad sense that includes both the cognitive structures and cognitive processes required to perform product-related tasks successfully. First, cognitive structure is the way people organize information [21]. The more detailed the cognitive structure is, the better a person would discriminate between information units. Since consumers with higher level of expertise possess more comprehensive cognitive structures, they usually have more detailed ideas about the product or product category. Attribute knowledge is one of such structures. Expert consumers should have more refined, complete and veridical knowledge of the product attributes [20]. But for novice consumers, many scholarly works have shown that they could not actively organize effective information search to evaluate the product, and that they can only passively perceive the product from external environment [22].

Second, cognitive processes are a series of tasks that an individual does continuously in a specific context [23]. They are various and dependent on the context, such as to identify, categorize objects, compare or discriminate among them, make preference judgments and so on [23]. In a shopping context, cognitive processes could be applying certain decision rules for acting on their product knowledge, such as analyzing information, identifying and isolating what is important and relevant and so on. Among them, comparative processes are commonly used for evaluation [24]. As noted in prior research, human judgment is comparative in nature [25], comparative processes took place in human perception [26], attitude formation [27] as well as decision making [24]. Comparisons do not exist in vacuum, but they are generated on the ground of acquisition and use of knowledge [25]. While consumers with less expertise can barely commence the comparative processes when evaluating a product, professionals or expert consumers are able to make comparative evaluation based on their product knowledge of attributes, as well as their knowledge of the specific market.

As professionals or expert consumers possess more highly developed conceptual structure of the product [20], the reviews would draw upon their extended knowledge structure and deliver more attribute-related thoughts and comparative evaluation of alternative products to equip others' understanding towards the product [28]. With the above discussions, we propose that expert reviews are characterized by two elements. The first is the information of product attributes' structure (PAS), which is defined as the overall structure of the product's important attributes. A complete PAS indicates a comprehensive and consumer knowledge structure delivered by the reviewer. The second element is the information of attributes' comparative evaluation (ACE), which is the attributes' comparative evaluation among different alternative products. The more detailed the comparison evaluation information is, the more cognitive processes and market knowledge of the reviewer is presented in the content.

Evaluating new products using online product reviews is increasingly common for consumers nowadays. It was reported that online shoppers are buying more of their purchases online rather than in stores, and that among them, most people make their product search online as well [29]. That is to say, a great number of consumers now acquire product information and construct their preferences in online information environment. Furthermore, online reviews, especially expert reviews, as one of the most important information sources, are of great influence for people to evaluate the product and make decisions [3, 5]. Therefore, given the situation, it is vital to develop understandings of how online reviews would influence consumers' product preferences and information consumption.

2.3 Preference Construction

Product attributes are at the center of product evaluation and preference formation [30]. A product could be assessed with a variety of attributes. Consumers might have their distinct needs and preferences for the various attributes. The constructive view of preference adheres to two tenets, (1) that the expressions of preference are constructed at the time the valuation is required, and (2) that the construction process will be shaped by the interaction between properties of human information processing system and properties of the decision task. [2, 31]. Many research has discussed the effects of choice and process variations on preference construction, concluding preferences that are highly labile and sensitive to contextual changes [2, 32].

Attribute preference construction could occur through learning about the various product information posted in the corresponding decision environment, especially when people are unfamiliar and inexperienced with the object [2]. This is because consumers may lack the cognitive resources to generated well-defined preferences, or because consumers bring multiple goals to a complex decision problem [31].

Existing studies have investigated the role of advertisements [33], recommendation agents [31] and online reviews [34]. One of their major findings is that when information about the attributes' importance is ambiguous, inclusion and highlight of a specific product attribute would affect the perceived importance of the attribute.

In the light of the numerous empirical evidence that the information environment may play an essential role in individuals' preference construction for multi-attribute products, we in this study, investigate the impact of expert reviews on consumers' attribute preference construction. In particular, we posit that expert reviews may influence the relative importance weight that consumers give to different attributes via the impact of the two types of embedded information, i.e. PAS information and ACE information.

This could be due to several mechanisms. First, for information processing with high involvement, the quality of information has a greater impact on persuasion [10]. As PAS provides a structural presentation of product attributes and ACE affords a horizontal comparison of attributes' performances among different products, more information cues (e.g. concreteness and informativeness) would be perceived by consumers [35, 36]. Therefore, expert reviews would be perceived as of high quality, which may lead to higher persuasive impact to novice consumers on product evaluation.

Second, the structural attribute information of product assessment supports consumers to learn from the expert reviews. When a novice consumer is evaluating a multi-attribute product, she would have a high motivation to know about the criteria that might be used to evaluate the product. The PAS and ACE content in expert reviews fits her needs at the spot and assists her learning of the product type. As the consumer's knowledge about the product increases, a higher confidence and efficacy to use the information would be yielded [37, 38], which in turn, increases the relative weight that consumer attach to the attributes in expert reviews.

Third, the product attribute-related information included in expert reviews could influence consumers' short-term memory. The attributes being discussed in the expert reviews would be more salient temporarily, hence heightening the perceived importance weight a consumer attaches to those attributes [31, 39]. Therefore, we hypothesize that,

H1. Expert reviews affect consumers by shaping their perceived importance of product attributes.

More specifically,

H1a. Product attributes structure information influences consumers' perceived importance of the attributes.
H1b. Comparative information of product attributes influences consumers' perceived importance of the attributes.

2.4   Peer-Review Consumption

Expert reviews are not the only information source that an online shopper has. As stated in prior literature, online reviews generated by peer consumers are influential in consumers' preference construction and decision-making process [34, 40]. An important follow-up question is whether and how consumers' early exposure to expert reviews would influence the absorption of peer reviews. Given that peer reviews could sway consumers' preferences [34], if expert reviews overpower the peer reviews, marketers or platform managers will be able to promote and take the advantages of content from experts. So, besides affecting consumers' preference construction, early exposure to expert reviews is also expected to influence consumers' consumption of peer reviews.

As mentioned, information about attributes structure and comparative evaluation in expert reviews could be perceived of higher quality [35], which encourages higher level of information use by novice consumers [41]. With their increasing knowledge, consumers' confidence in the information source would also increase. According to dissonance theory, a human is not motivated to be right, instead, s/he is motivated to believe that s/he is right [42]. Thus, their confidence in experts may prevent them from learning extra information, leading to neglect and reduction of further information acquisition. Therefore, we hypothesize that,

H2. When consumers are pre-exposed to expert reviews, their consumption of peer review information would be reduced.

3   Proposed Methodology

We intend to utilize an experiment to directly manipulate expert review content to test our hypotheses. We create our experimental conditions in a 2 (complete vs. incomplete PAS information) by 2 (detailed vs. brief ACE information) between-subjects design. Participants will be given a purchase scenario and a piece of manipulated expert review. Then they will be asked to respond to our questions before and after we show them the peer reviews. As a control, we also include a condition where no expert review is shown to participants.

4   Potential Contribution and Limitations

4.1   Potential Contribution

First, our research proposed two elements to characterize the expert reviews from the lens of profession research. Our findings have the potential of enlightening content providers to generate more powerful content and marketers to make use of reviews and reform their marketing strategies.

Second, our research is among the first studying the impact of expert reviews on consumers' subsequent information consumption of peer reviews. Our paper will have both theoretical and practical implications on consumer behaviors in online markets. The results on consumer information consumption will further extend the research findings that expert reviews may influence the swaying effects brought by peer reviews.

Last but not least, our findings will lead to a better understanding of the differences between expert reviews and peer reviews. While a large number of studies examining online reviews exist, very few papers discussed their differences. By drawing on the profession research, our theoretical development and analysis on review content would provide another perspective of distinguishing the two types of online reviews.

4.2   Limitations

Our proposal has several limitations. First, when defining expert reviews, we only use comparison information as a representative of experts' cognitive processes delivered in an expert review. Other aspects, such as using logic, categorization and excluding alternatives may also be included. We call for future studies targeting on other content of cognitive processes presented in expert reviews. Second, we do not consider the impact of irrelevant attributes in our experiment. As we focus on the impact of expert reviews, which would rarely contain information about irrelevant attributes' information, we think it is more reasonable to avoid any distraction from those attributes. The third limitation is the arrangement of consumers' reading order. In our study, we are to investigate the impact of early-exposure of expert reviews. However, in practices, peer reviews, instead of expert reviews, may be exposed to consumers first when they are looking for the product information. We argue that as a first study to investigate the expert reviews on the subsequent information acquisition, our findings would build the basis for and shed light upon future studies to examine the topic with more depth and variation.

Acknowledgement. This research is supported in part by the General Research Fund from the Hong Kong Research Grants Council (#17514516B) and the Seed Funding for Basic Research from the University of Hong Kong (#104003314).

References

1. Cordell, V.V.: Consumer knowledge measures as predictors in product evaluation. Psychol. Mark. **14**(3), 241–260 (1997)
2. Bettman, J.R., Luce, M.F., Payne, J.W.: Constructive consumer choice processes. J. Consum. Res. **25**(3), 187–217 (1998)
3. Dellarocas, C.: The digitization of word of mouth: promise and challenges of online feedback mechanisms. Manag. Sci. **49**(10), 1407–1424 (2003)
4. Dellarocas, C., Awad, N., Zhang, M.: Using online ratings as a proxy of word-of-mouth in motion picture revenue forecasting. Citeseer (2005)
5. Luo, X., Gu, B., Zhang, J., Phang, C.W.: Expert blogs and consumer perceptions of competing brands. MIS Q. **41**(2), 371–395 (2017)
6. Fred, S., Examining endorsement and viewership effects on the source credibility of YouTubers. University of South Florida (2015)
7. Zhou, W., Duan, W.: Do professional reviews affect online user choices through user reviews? an empirical study. J. Manage. Inform. Syst. **33**(1), 202–228 (2016)
8. Amblee, N., Bui., T.: Freeware downloads: an empirical investigation into the impact of expert and user reviews on demand for digital goods. In: Proceedings of AMCIS 2007 (2007)
9. Lawrence, B., Fournier, S., Brunel, F.: When companies don't make the ad: a multimethod inquiry into the differential effectiveness of consumer-generated advertising. J. Advertising **42**(4), 292–307 (2013)
10. Petty, R.E., Cacioppo, J.T., Goldman, R.: Personal involvement as a determinant of argument-based persuasion. J. Pers. Soc. Psychol. **41**(5), 847 (1981)

11. Mantel, S.P., Kardes, F.R.: The role of direction of comparison, attribute-based processing, and attitude-based processing in consumer preference. J. Consum. Res. **25**(4), 335–352 (1999)

12. Bao, Z., Chau, M.: The impact of the collective rating presence on consumers' perception. In: Proceedings of the International Conference on Electronic Business (ICEB 2015). Hong Kong (2015)

13. Clarke, K., Belk, R.W.: The effects of product involvement and task definition on anticipated consumer effort. Adv. Consum. Res. **6**, 313–318 (1979)

14. Gu, B., Park, J., Konana, P.: Research note—the impact of external word-of-mouth sources on retailer sales of high-involvement products. Inf. Syst. Res. **23**(1), 182–196 (2012)

15. Cheung, C.M.-Y., Sia, C.-L., Kuan, K.K.: Is this review believable? a study of factors affecting the credibility of online consumer reviews from an ELM perspective. J. Assoc. Inf. Syst. **13**(8), 618 (2012)

16. Huang, J.H., Chen, Y.F.: Herding in online product choice. Psychol. Mark. **23**(5), 413–428 (2006)

17. Li, M., Huang, L., Tan, C.-H., Wei, K.-K.: Helpfulness of online product reviews as seen by consumers: source and content features. Int. J. Electron. Commer. **17**(4), 101–136 (2013)

18. McGaghie, W.C.: Professional competence evaluation. Educ. Res. **20**(1), 3–9 (1991)

19. Freidson, E.: Professional Powers: A Study of the Institutionalization of Formal Knowledge. University of Chicago Press, Chicago (1988)

20. Alba, J.W., Hutchinson, J.W.: Dimensions of consumer expertise. J. Consum. Res. **13**(4), 411–454 (1987)

21. De Bont, C.J., Schoormans, J.P.: The effects of product expertise on consumer evaluations of new-product concepts. J. Econ. Psychol. **16**(4), 599–615 (1995)

22. Kahn, B.E., Meyer, R.J.: Consumer multiattribute judgments under attribute-weight uncertainty. J. Consum. Res. **17**(4), 508–522 (1991)

23. Nosofsky, R.M.: Similarity scaling and cognitive process models. Annu. Rev. Psychol. **43**(1), 25–53 (1992)

24. Kahneman, D., Miller, D.T.: Norm theory: comparing reality to its alternatives. Psychol. Rev. **93**(2), 136 (1986)

25. Mussweiler, T.: Comparison processes in social judgment: mechanisms and consequences. Psychol. Rev. **110**(3), 472 (2003)

26. Bao, Z., Chau, M.: The effect of collective rating on the perception of online reviews. In: 20th Pacific Asia Conference on Information Systems (2016)

27. Vallone, R.P., Ross, L., Lepper, M.R.: The hostile media phenomenon: biased perception and perceptions of media bias in coverage of the Beirut massacre. J. Pers. Soc. Psychol. **49**(3), 577 (1985)

28. Connors, L., Mudambi, S.M, Schuff, D.: Is it the review or the reviewer? a multi-method approach to determine the antecedents of online review helpfulness. In: 2011 Proceedings of 44th Hawaii International Conference on System Sciences (HICSS). IEEE (2011)

29. Stevens, L. Survey Shows Rapid Growth in Online Shopping. Surveyed shoppers made 51% of their purchases on the web 2016, 8 June 2016. https://www.wsj.com/articles/survey-shows-rapid-growth-in-online-shopping-1465358582. Accessed 1 May 2018

30. Lancaster, K.: Consumer Demand: A New Approach. Columbia University Press, New York (1971)

31. Häubl, G., Murray, K.B.: Preference construction and persistence in digital marketplaces: the role of electronic recommendation agents. J. Consum. Psychol. **13**(1–2), 75–91 (2003)

32. Tversky, A., Kahneman, D.: Rational choice and the framing of decisions. J. Bus. **59**(4), S251–S278 (1986)

33. Gardner, M.P.: Advertising effects on attributes recalled and criteria used for brand evaluations. J. Consum. Res. **10**(3), 310–318 (1983)
34. Liu, Q.B., Karahanna, E.: The dark side of reviews: the swaying effects of online product reviews on attribute preference construction. MIS Q. **41**(2), 427–448 (2017)
35. Harmon, R.R., Razzouk, N.Y., Stern, B.L.: The information content of comparative magazine advertisements. J. Advertising **12**(4), 10–19 (1983)
36. Bhattacherjee, A., Sanford, C.: Influence processes for information technology acceptance: an elaboration likelihood model. MIS Q. **30**(4), 805–825 (2006)
37. Gist, M.E., Mitchell, T.R.: Self-efficacy: a theoretical analysis of its determinants and malleability. Acad. Manag. Rev. **17**(2), 183–211 (1992)
38. Murray, K.B.: A test of services marketing theory: consumer information acquisition activities. J. Mark. **55**(1), 10–25 (1991)
39. Herr, P.M.: Priming price: prior knowledge and context effects. J. Consum. Res. **16**(1), 67–75 (1989)
40. Zhang, K.Z., Zhao, S.J., Cheung, C.M., Lee, M.K.: Examining the influence of online reviews on consumers' decision-making: a heuristic–systematic model. Decis. Support Syst. **67**, 78–89 (2014)
41. O'Reilly, C.A.: Variations in decision makers' use of information sources: the impact of quality and accessibility of information. Acad. Manag. J. **25**(4), 756–771 (1982)
42. Bell, J.: The effect of presentation form on judgment confidence in performance evaluation. J. Bus. Finance Acc. **11**(3), 327–346 (1984)

FinTech

An Empirical Investigation of Equity-Based Crowdfunding Campaigns in the United States

Son Bui[1][✉] and Quang "Neo" Bui[2]

[1] Texas A&M University Commerce,
2200 Campbell Street, Commerce, TX 75428, USA
son.bui@tamuc.edu
[2] Rochester Institute of Technology,
105 Lomb Memorial Drive, Rochester, NY 14623, USA
qnbbbu@rit.edu

Abstract. Following the introduction of the JOBS Act in 2016, equity-based crowdfunding has become an alternative e-Business model for startups to fund their companies. Since then, the number of platforms that offer equity-based crowdfunding as well as the total investment in equity-based crowdfunding has steadily increased. Yet, empirical research on equity-based crowdfunding has been lagging, and the empirical evidence has suggested some inconsistent findings across different contexts. Against this backdrop, this paper investigates the success factors for equity-based crowdfunding campaigns. Using a dataset collected from the EquityNet and CrunchBase platforms, we find that lack of prior experience with fundraising is the most important factor that helps equity-based crowdfunding campaigns attract any capital at all from investors; while the number of social networking connections of the core management team and the company valuation will determine the amount of capital that a business can raise through equity-based crowdfunding. Our findings call for additional research that looks at success factors for different types of outcomes in equity-based crowdfunding.

Keywords: Equity-based crowdfunding · Startup

1 Introduction

Equity-based crowdfunding (ECF) refers to the process of fundraising through Internet-based platforms and by offering equity stakes to investors in exchange for capital [2, 16]. In recent years, across countries, ECF has increasingly become an important fundraising means for startups, especially technology firms, to obtain capital from the general public [6, 11, 15]. In the United States, title II of the JOBS Act legalized ECF for accredited investors[1] in 2012; and in 2016, title III of the JOBS Act expanded the scope of ECF to the general public. Since then, the ECF market has steadily grown to be one of the stable investment choices besides other types of investments such as

---

[1] Accredited investors are individuals who either have more than $200,000 in income per year or have at least $1 million in assets.

© Springer Nature Switzerland AG 2019
J. J. Xu et al. (Eds.): WEB 2018, LNBIP 357, pp. 111–123, 2019.
https://doi.org/10.1007/978-3-030-22784-5_11

traditional venture capital or other types of crowdfunding (e.g., reward-based crowdfunding). Within a year, total ECF investment in the US grew from $27 million in 2016 to $76 million in 2017, with a projection of up to $1 billion in the next five years [10].

There are several factors that make it important to study ECF campaigns. First, compared to other types of investment, ECF has some distinctive characteristics. Unlike investors in traditional fundraising methods (e.g., venture capital), ECF investors typically are less knowledgeable and conduct less due diligence in assessing investment opportunities [1, 22]. Compared to other types of crowdfunding, ECF investors are motivated by equity stakes rather than by products (reward-based crowdfunding) or interest payments (loan-based crowdfunding) [16]. Second, empirical findings about ECF success vary across different contexts. For example, Ahlers et al. [2] found that social capital factors had no impact on funding success for ECF campaigns, while Vismara [23] found that social capital factors had significant impact on success for ECF campaigns. Yet, to date, while the number of empirical studies of ECF is growing, the findings are inconsistent in IS literature.

Against this backdrop, it is important to conduct additional empirical research on ECF to understand what contributes to the success of ECF campaigns. In this paper, we investigate the research question: *What are the success factors for ECF campaigns?* Drawing from prior studies, we use social networking theory and signal theory to study ECF success factors. We examine a dataset of 99 ECF campaigns in the EquityNet platform, one of the leading ECF platforms in the US. The findings suggest that ventures are likely to raise *some* capital from ECF if they have had less prior experience with fundraising campaigns. In addition, among ventures that have raised capital from ECF, their social networking connections and financial valuation can significantly increase the amount of capital that they can raise through ECF.

This paper makes several contributions to the ECF research. First, we distinguish different possible outcomes for ECF success and find that different signals are associated with different outcomes. In light of prior studies which suggest that there are intangible outcomes for ECF such as company valuation and validation [5], our findings call for additional research on success factors for a range of other possible ECF outcomes (e.g., raising *some* but not *all* capital, follow-on funding, social and intangible benefits). Secondly, our findings suggest the possibility of investigating the development stages that lead to the adoption of ECF as a strategy for fundraising. That is, how and why entrepreneurs decide to take on ECF as opposed to other types of fund-raising approaches. To date, this has only received limited attention from researchers.

The rest of the paper is organized as follows. We first provide a background on crowdfunding and ECF, then develop a research model and research hypotheses based on prior studies. Next, we present our method and findings of the study. We conclude with discussions of the findings regarding current theory and practice of ECF campaigns.

2   Background on Equity-Based Crowdfunding

Crowdfunding is an umbrella term that describes a form of fundraising through Internet-based platforms, whereby a group of people pool money in individual contributions to support a particular goal [2]. Compared to other traditional methods of fundraising such as venture capital or business angels, crowdfunding typically attracts unsophisticated investors, many of whom have limited investment experience but seek alternative ways to leverage their capital [1]. In fact, studies of crowdfunding investors in the UK show that most investors have a small investment portfolio (below £5,500) that comes mostly from their savings rather than investment budgets [22]. Due to their lack of experience and capability, crowdfunding investors often conduct less due diligence in studying investment opportunities, while the traditional fundraising methods often involve extensive due diligence processes (e.g., face-to-face interactions, multiple rounds of presentations) before decisions are made [2, 22, 23]. This leads to the classic information asymmetry problem which often prompts crowdfunding investors to exhibit herding behaviors and rely on crowd wisdom in their decision making [1].

Table 1. Comparing different forms of fundraising

	Crowdfunding				Venture capital, business angels
	Equity-based	Reward-based	Loan-based	Donation-based	
Typical investor	Mixed, many have limited investment experience				Qualified institutions or individuals
Due diligence	Limited, conducted by individuals				Extensive
Investment amount	Small to medium (hundreds to tens of thousands)				Medium to large (up to millions)
Deal flow	Through Internet-based platforms				Through face-to-face interactions
Motivations	Equity	Product or service	Earned of interests	Altruism	Equity
Risks	Loss of investment	Delivery failure	Loss of principle	None	Loss of investment
Example	EquityNet	Kickstarter	Lending Club	GoFundMe	Benchmark Capital

In general, there are four types of crowdfunding: *donation-based, loan-based, reward-based*, and *equity-based* [4, 14–16]. They are different in terms of motivations and risks (see Table 1). In *donation-based* crowdfunding, investors have altruistic motivations to donate charitable contributions in support of good causes (e.g., paying for an expensive surgery in GoFundMe) [15]. In these fundraising campaigns, investors do not expect monetary returns and often find satisfaction in supporting campaigns that resonate with their intrinsic values [7, 8] or those that advance a specific social cause [14]. *Loan-based* or debt-based crowdfunding offers peer-to-peer lending opportunities in which a group of lenders would pool money together as a loan to individuals or businesses with the expectation that the loan will be paid off together with the interests added [16]. It accounts for the largest amount of total crowdfunding volume in 2014, totaling up to 68% of global market share [14]. Unlike donation-based crowdfunding, which expects no return and suffers little risk, loan-based crowdfunding investors expect a small return on investment in the form of interest paid on the original loan while incurring a risk of losing the principal amount in the event of default by borrowers.

Reward-based crowdfunding offers non-monetary rewards to investors, either through products or services. Thus, reward-based campaigns often attract early innovation adopters who are motivated by the access to new and sophisticated gadgets that are not yet available to the public [15, 16]. Because this type of fundraising often relies on the ability of entrepreneurs to deliver new and innovative products or services, it is highly receptive to the risk of fraud or the incompetency of the entrepreneurs to deliver their promises [1]. *Equity-based* crowdfunding, on the other hand, offers equity share of a business in exchange for contributions. In this regard, it is similar to traditional fundraising methods because investors are incentivized by equity shares in the target business [15]. However, unlike the case with traditional methods, equity-based crowdfunding entrepreneurs disclose their information on Internet-based platforms instead of through face-to-face interactions, and have limited opportunity to defend their campaigns through outside assistance such as reputations of intermediaries and financial analysts [23].

In this paper, we are particularly interested in equity-based crowdfunding and its success factors for several reasons. First, as a phenomenon, equity-based crowdfunding is relatively new and has distinctive characteristics compared to the other types of fundraising methods. Equity-based crowdfunding investors are usually young and inexperienced individuals who lack the due diligence in examining investment opportunities [22]. As a result, they rely heavily on social clues and crowd due diligence to assist their decision making [1, 2]. In addition, because equity-based crowdfunding investors are incentivized by equity shares rather than by sophisticated and new products (rewarded-based crowdfunding), or by monthly interest payments (loan-based crowdfunding), it is likely that their decision making is informed by different criteria than other types of crowdfunding. Indeed, several studies have highlighted various success factors for equity-based crowdfunding success that depart from traditional factors found in venture capital investments or by other types of crowdfunding campaigns [2, 4, 7, 8, 15, 16, 23].

Second, from the theoretical perspective, empirical studies on equity-based crowdfunding are lagging compared to other streams of fundraising research. Because equity-based crowdfunding was only recently made legal in the US, with the enactment of Title II of the JOBS Act in 2012 (for accredited investors) and Title III of the JOBS Act in 2016 (for the general public), the amount of research on equity-based crowdfunding is still limited. In addition, prior studies hint at some inconsistent findings across different contexts. Some suggest that social capital from networks and business linkages has a positive impact on the success of an equity-based crowdfunding campaign [15, 23], while others show that social capital has little to no impact on crowdfunding success [2]. Moreover, both Ahlers et al. [2] and Mamonov and Malaga [16] found that intellectual capital—measured by the number of patent holdings—has no impact on crowdfunding success; while Ralcheva and Roosenboom [21], in agreement with studies of venture capital firms, found that patent holding can significantly increase the chance of success for equity-based crowdfunding campaigns.

In sum, the distinctive characteristics of ECF, the lack of empirical studies in IS literature, and the inconsistent findings present a research opportunity to further investigate the success factors of equity-based crowdfunding. Next, we develop our research model.

3 Research Model and Hypotheses Development

Drawing from the extant literature, and given that ECF often draws inexperienced investors who are susceptible to social influences, we argue that campaign character-istics and social signals can increase the chance of crowdfunding success. Our research model is presented in Fig. 1.

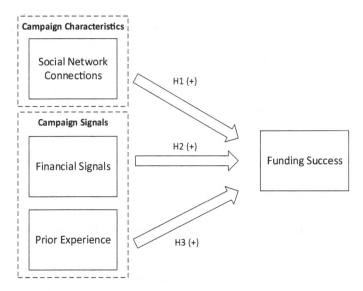

Fig. 1. Research model

Crowdfunding success is a multifaceted concept, and prior research has considered a wide range of possible success measures such as whether a campaign is fully funded, whether the campaign raised the minimum amount of capital that was sought, the amount of capital raised, the number of investors, and speed of investment [2, 15, 16, 23]. In this study, we focus on two success measures: the amount of capital raised, and whether a campaign was able to raise *any* capital. By targeting these two success measures, we focus on crowdfunding platforms that allow entrepreneurs to keep any amount of the capital that they raised (i.e., flexible funding) instead of platforms that require entrepreneurs to meet their goal to gain access to the raised capital (i.e., all-or-nothing funding). This mechanism is called *provision point* which is designed to reduce coordination and free-riding problems in crowdfunding campaigns [1, 6]. By focusing on campaigns without a provision point mechanism, we can examine how other factors influence investors' decisions. In addition, Brown et al. [5] suggest that there are various intangible benefits of crowdfunding. Nearly all respondents from among 42 UK entrepreneurs who have successfully obtained capital through ECF acknowledged intangible benefits such as access to new customers, media and press attention, and validation of their products/services and business model. In other words, gaining any amount of capital through crowdfunding can potentially provide additional benefits to the entrepreneurs. Thus, our success measures allow us to be more sensitive to situations when a campaign does not meet its goal (i.e., does not meet the provision point) but is still able to gain some capital and therefore gain intangible benefits through ECF.

Prior studies suggest that campaign characteristics can determine campaign success [2, 13, 16]. Because ECF campaigns target the general public, the role of a business's social network capital has been found to be a significant success driver [15, 18, 23]. The social network capital refers to the strengths that come from the social connections and networks of a business's management team (e.g., LinkedIn network, MBA graduate network). Social network connections help a business spread information, generate worth-of-mouth, and solicit early contributions that jumpstart the campaign. This is especially true for crowdfunding campaigns in which many investors lack due diligence and rely on social clues and crowd due diligence to augment their decisions [1]. Among other types of crowdfunding, studies have found evidence of social capital in early contributions in reward-based crowdfunding campaigns [9, 18, 19], or social network effects and activity in donation-based crowdfunding campaigns [20]. Thus, we propose:

Hypothesis 1: *A business's social network connections are positively associated with (a) the probability of raising capital, and (b) the amount of capital raised through equity-based crowdfunding.*

Prior crowdfunding research has examined the various signals that a business can include in their crowdfunding campaign to reduce information asymmetries and uncertainty for potential investors [1, 2, 18, 19]. Because a majority of ECF investors are inexperienced and lack due diligence in accessing investment opportunities [1, 17, 22], the more effective signals that a business can provide, the more likely that the business can successfully raise capital through crowdfunding campaigns [2]. Of the many signals, financial signals can be a clear indicator of success as they directly communicate to investors how the business conducts itself financially, and whether the business is projected to succeed in the future [16]. These financial signals are especially important for ECF campaigns in which investors are incentivized by financial motives [5, 8]. Thus, we suggest:

Hypothesis 2: *A business's financial signal is positively associated with (a) the probability of raising capital, and (b) the amount of capital raised through equity-based crowdfunding.*

In addition, the level of uncertainty of a business will impact the likelihood that an investor will invest in a new venture. Given that ECF is a new type of investment, many investors will use existing information to ascertain the likelihood for success of the company. Prior studies have found that previous success with fundraising campaigns is a strong success factor for crowdfunding campaigns [16]. This is especially true for ECF when many investors are non-professionals and inexperienced, and rely on easy-to-understand information to determine the likelihood for success of a company [1, 2, 23]. Prior success will lower uncertainty for these investors and assure them of future success. In addition, entrepreneurs who have prior experience with fundraising will be able to apply lessons learned from their experience and thus will be likely to avoid common mistakes and make their projects more appealing to investors. Thus, we suggest:

Hypothesis 3: *A business's previous experience with fundraising campaigns is positively associated with (a) the probability of raising capital, and (b) the amount of capital raised through equity-based crowdfunding.*

4   Methods

4.1   Data Collection

We collected data from multiple existing databases including CrunchBase and EquityNet for this study. CrunchBase is the largest public database of private startup companies, containing information on startups' founders, products, funding, investors, news, board of directors, and top managers, among others. The second database, EquityNet, is one of the top crowdfunding platforms for entrepreneurs to raise money, and for investors to find potential startups to invest money. However, unlike similar platforms such as Indiegogo or Kickstarter, EquityNet is less restricted and more flexible for entrepreneurs to create an ECF plan.

First, we used the CrunchBase database to obtain the list of US startup companies that have used ECF as one of their fund-raising campaigns from the launch of the database to May 1st 2017. Our data consist of 2,657 US private startups that have at least one ECF activity in their fund-raising history. For each startup, we collected data about fundraising experience, startup maturity, top management team members, and board members. Next, we used publicly available information on the EquityNet database to cross check our data to make sure that the startups have fund-raising campaigns listed in the EquityNet database, and collected data on each startup's ECF campaigns. Detailed information about ECF campaigns such as amount of capital raised, campaign funding goal, startup valuation, and popularity rating has been gathered from EquityNet. After verifying and eliminating all missing data, our data were reduced to 454 startups that have information about each company's profile and ECF campaign.

Second, for each startup, to collect social network connection data, we tracked the LinkedIn connections (i.e., number of followers) for each member of the top management team, and board and advisor teams. We then averaged the LinkedIn connections to calculate the social network connections for each startup. During our data collection from LinkedIn, all profiles that have hidden data or an arbitrary number of followers (e.g., 500+) are excluded from our calculation. After filtering out all missing data from CrunchBase, EquityNet, and LinkedIn, our final sample data include 99 companies with a total of 343 profiles of top management team members and board of advisor team members that have been collected from LinkedIn. The variables are described and explained in the following subsections.

4.2   Dependent and Independent Variables

To measure ECF success, we use two variables, Campaign Funded and Amount of Capital Raised, as our main dependent variables.

Campaign Funded. We used the probability of whether ECF raised any funding as one proxy for ECF success. If startups successfully raised any money during their campaign, the campaign funded has a value of 1, and 0 otherwise.

Amount of Capital Raised. The second proxy of ECF success was operationalized by the total amount of funding raised during the ECF campaign.

Our primary independent variables include fundraising experience, startup valuation, and social network connections. These measures are based on equity campaign characteristics and equity campaign signals.

Fundraising Experience. We operationalized startups' previous experience with fundraising by capturing the total number of fundraising rounds that the startups have had prior to their ECF campaigns. Within our dataset, no company has had more than one round of ECF. A higher number of fundraising experiences means that the startups have received funding from previous funding rounds such as seed rounds, series A, series B, series C, and so on. A '0' value of fundraising experience means that the startups use ECF to fund their seed round.

Startup Valuation. We measured each startup's financial signal by capturing its company valuation, the log of the current value of the startup as evaluated by the founder(s) during the ECF campaign. The valuation of startups is submitted by the founder(s) as one of the requirements to raise funding in EquityNet.

Social Network Connections. We operationalized startups' social network connections by capturing the log of the average of LinkedIn connections of startups' top management team, and board and advisor team. This measure is based on external connections outside of the startups upon the nodes of the top management team, and board of advisor team. The higher the number, the more external connections startups have.

4.3   Control Variables

We use popularity rating and startup maturity to control for campaign heterogeneity.

Popularity Rating. This number shows the level of interest that the startup has received from the public compared to other peer startups. These data are calculated by EquityNet based on how many times the equity campaign documents were viewed and downloaded, how fast the startups responded to inquiries, and the amount of funding raised compared to the funding goal.

Startup Maturity. This number shows the total number of years from a startup's founding date to the ECF campaign's starting date. A higher number means the startup began an ECF campaign later in their funding lifecycle (e.g., series A, series B etc.). A negative number means the startups began an ECF campaign before the startup's official founding date (Table 2).

Table 2. Descriptive statistics

	N	Mean	SD	Min	Max
Dependent variables					
Amount of equity raised (in dollar)	63	607,272	1,195,065	1,000	5,600,000
Campaign funded (binary)	99	0.63	0.54	0	1
Independent variables					
Social network connection (log value)	99	3.28	3.50	0.30	4.21
Startup evaluation (log of dollar)	99	6.92	7.18	0.60	7.93
Funding raising experience	99	1.50	0.97	1	6
Control variables					
Popularity rating	99	3.71	0.58	0.1	5
Firm maturity (in years)	99	3.23	4.15	−0.98	32.08

5   Results

Because our dependent variable for H1a, H2a, and H3a is binary, we performed a standard logit regression analysis for all 99 startups in our final sample data. Out model can be written as:

$$logit(CampaignFunded) = \alpha_0 + \alpha_1 SocialNetworkConnection + \alpha_2 StartupValuation$$
$$+ \alpha_3 FundRaisingExperience + \alpha_4 PopularityRating$$
$$+ \alpha_5 StartupMaturity$$

To test for our H1b, H2b, and H3b, we performed a multilinear regression analysis only for startups that received funding. Out of 99 startups, 63 firms are used for the second model. Our multilinear regression model can be written as:

$$AmountofCapitalRaised = \alpha_0 + \alpha_1 SocialNetworkConnection + \alpha_2 StartupValuation$$
$$+ \alpha_3 FundRaisingExperience + \alpha_4 PopularityRating$$
$$+ \alpha_5 StartupMaturity$$

Table 3 presents the parameter estimates of our models. Overall, model 1 is significant (likelihood ratio $\chi^2 = 14.89$). The explanatory power of the logit model is 0.1127. From model 1, social network connections and startup valuation have an insignificant impact on an equity campaign being funded ($\alpha_1 = 0.013, p > 0.05$; $\alpha_2 = 0.086, p > 0.05$). Thus, H1a and H2a are unsupported. On the other hand, the results suggest that fundraising experience has a significant negative impact on whether the campaign is funded ($\alpha_3 = -0.957, p < 0.01$). In other words, the less fundraising experience startups have, the more likely they can raise capital through ECF. It means that our H3a is partially supported.

In our model 2, the explanatory power of the multilinear regression is 0.4672. Both social network connections and startup valuation have significant positive impact on the amount of capital raised ($\alpha_1 = 0.242, p < 0.05$; $\alpha_2 = 0.485, p < 0.01$). Thus, H1b and H2b are supported. Our results also suggest that fundraising experience has insignificant impact on amount of capital raised ($\alpha_3 = 0.235, p > 0.05$). It means that our H3b is unsupported.

Table 3. Model results

	Model 1 Logit (campaign funded)	Model 2 Amount of capital raised
Intercept	3.13	0.93
Social network connections	0.013	0.242*
Startup valuation	0.086	0.485***
Funding raising experience	−0.957**	0.235
Popularity rating	−0.603	0.541
Startup maturity	−0.0002	0.032
Sample size	99	63
Log likelihood χ^2	14.63	
R2	0.1127	0.4672
Adjusted R2	0.0742	0.4205

*p < 0.05, **p < 0.01, ***p < 0.001

6 Discussion

Given the rapid development of ECF, increasingly research has examined success factors of ECF campaigns. This study contributes to the growing body of ECF literature by examining how social network connections, startup valuation, and fundraising experience affects ECF. The findings offer several contributions to theories and practices of ECF.

First, we examined ECF success by measuring whether the campaign is funded *at all* as well as the amount of capital the campaign can raise. We showed that how different factors of ECF campaigns, that is, ECF campaign characteristics and ECF campaign signals, affect these success outcomes differently. Our results indicate that in order to receive any funding, startups that had less traditional funding rounds are more likely to receive funding through ECF. Social network connections and startup valuation, on the other hand, have no impact on whether a startup receives funding from ECF. This finding is couterintuitive and contradicts previous studies [16] which suggested that previous crowdfunding experience has a positive impact on future funding because it is an indicator of startups' high performance. However, our finding suggests that in the US market where ECF is a relatively new and disruptive phenomenon for startups to raise capital, having too much experience in traditional funding rounds can be a negative factor. Because most investors of ECF are individuals rather than organizations, they tend to be non-professional and inexperienced in evaluating a new crowdfunding method such as ECF [1, 2, 23]. Thus, they are often looking for a short-term investment with high reward and high risk—a phenomenon that is known as the expected utility hypothesis in economics research [3, 12]. According to this hypothesis, individuals can make "irrational" choices with high uncertainty and high risks to maximize potential returns. Thus, startups with less experience in fundraising are considered as potential targets for ECF investors due to the odds of high returns in spite of the associated high risks.

Second, we extend our research question to examine what factors contributed to the amount of capital raised through ECF. This question is particularly important given a number of ECF platforms use a flexible funding model without a provision mechanism point [1, 6]. In situations in which startups can keep all capital raised, our results suggest that social network connections and startup valuation are indeed significant predictors of the amount of capital raised through ECF—a striking difference from the previous outcome. This implies that to reach their fundraising goal, startups need to have strong social connections with external investors, and the company has to show high potential for return investment back to investors. This aligns with previous studies of crowdfunding that find strong worth-of-mouth and high startup valuation are strong indicators of online funding campaign success in which many investors lack due diligence and rely on social clues and crowd wisdom to augment their decisions [1].

Finally, our findings together suggest that there are different success factors for different outcomes of an ECF campaign: to raise *any* capital at all, prior experience with traditional fundraising and the "newness" of the startup are important, while to increase the amount of capital raised, startups need to focus on their social network connections and valuation. In light of prior studies which suggest that there are intangible outcomes for ECF such as company valuation and validation [5], and that the provision point mechanism, which only allows fund withdrawal when a campaign goal is met, is critical to campaign success [1, 6], our findings call for additional research on success factors for other possible outcomes (e.g., raising *some* but not *all* capital, follow-on funding, social and intangible benefits). This also suggests a promising line of research on the development stages that lead to the adoption of ECF as a strategy for fundraising. That is, how and why entrepreneurs decide to take on ECF as opposed to other types of fundraising approaches.

6.1 Managerial Implications

Our findings offer several implications to ECF practices. First, our findings provide different strategies for startup managers to adopt ECF as a strategy for fundraising. Given that most investors for ECF are looking for high reward/high return startups to invest in, startups in early stages have a greater chance to receive funding than those in later stages and therefore should explore ECF as an alternative channel for seed funding. In addition, startups need to allocate resources to increase the external company network to outside investors and raise valuation through innovative products. Online ECF platforms such as EquityNet are excellent resources for startups to reach more capital, while other crowdfunding platforms such as Indiegogo or Kickstarter are more competitive nowadays to raise funding.

For platform designers, our findings suggest that platforms should pay attention to different enabling factors to help ECF campaigns. Particularly, platforms with a focus on all-or-nothing funding models should offer ways to allow startups to highlight and leverage social network connections and company valuation. Mechanisms such as quality signals, feedback systems, and trustworthy intermediaries can be of great value [1]. On the other hand, platforms that have a flexible funding model can implement features that highlight the novelty and innovativeness of the startups to attract investors. Prior studies have suggested mechanisms such as videos or quality of project description can be significant predictors of success [18].

The study is not without limitations. Our data are based on existing databases that are populated by self-report data from startups, entrepreneurs, and volunteers. Thus, there are limits to the data. Future studies are encouraged to duplicate our study, or combine it with additional data collection methods (e.g., survey, interviews) to enrich the insights suggested in this study.

7 Conclusion

In recent years, crowd markets have grown significantly and become a global phenomenon. Startups are increasingly turning to platforms such as Kickstarter to raise capital, bootstrap their customer base, and connect to potential investors. In the US market, ECF is still a relatively new fundraising channel as it was only legalized to the general public in 2016. To further understand this new phenomenon, this study examines success factors of ECF campaigns among US startups. Our findings suggest that depending on different funding outcomes, different factors will play different roles. This calls for further research to understand this phenomenon and how it impacts the design of ECF platforms as well as the structure of ECF campaigns.

References

1. Agrawal, A., Catalini, C., Goldfarb, A.: Some simple economics of crowdfunding. Innov. Policy Econ. **14**(1), 63–97 (2014)

2. Ahlers, G.K.C., Cumming, D., Günther, C., Schweizer, D.: Signaling in equity crowdfunding. Entrepreneurship Theor. Pract. **39**(4), 955–980 (2015)
3. Bell, D.E.: One-switch utility functions and a measure of risk. Manage. Sci. **34**(12), 1416–1424 (1988)
4. Belleflamme, P., Lambert, T., Schwienbacher, A.: Crowdfunding: tapping the right crowd. J. Bus. Ventur. **29**(5), 585–609 (2014)
5. Brown, R., Mawson, S., Rowe, A., Mason, C.: Working the crowd: improvisational entrepreneurship and equity crowdfunding in nascent entrepreneurial ventures. Int. Small Bus. J. Res. Entrepreneurship **36**(2), 169–193 (2018)
6. Burtch, G., Hong, Y., Liu, D.: The role of provision points in online crowdfunding. J. Manage. Inf. Syst. **35**(1), 117–144 (2018)
7. Cecere, G., Guel, F.L., Rochelandet, F.: Crowdfunding and social influence: an empirical investigation. Appl. Econ. **49**(57), 5802–5813 (2017)
8. Cholakova, M., Clarysse, B.: Does the possibility to make equity investments in crowdfunding projects crowd out reward-based investments? Entrepreneurship Theor. Pract. **39**(1), 145–172 (2015)
9. Colombo, M.G., Franzoni, C., Rossi-Lamastra, C.: Internal social capital and the attraction of early contributions in crowdfunding. Entrepreneurship Theor. Pract. **39**(1), 75–100 (2015)
10. Crowdfund capital advisors: The 2017 state of regulation crowdfunding. Crowdfund Capital Advisors (2018)
11. De Crescenzo, V.: The role of equity crowdfunding in financing SMEs: evidence from a sample of European platforms. In: Bottiglia, R., Pichler, F. (eds.) Crowdfunding for SMEs, pp. 159–183. Palgrave Macmillan, London (2016)
12. Friedman, M., Savage, L.J.: The expected-utility hypothesis and the measurability of utility. J. Polit. Econ. **60**(6), 463–474 (1952)
13. Hornuf, L., Neuenkirch, M.: Pricing shares in equity crowdfunding. Small Bus. Econ. **48**(4), 795–811 (2017)
14. Kim, H., Moor, L.: The case of crowdfunding in financial inclusion: a survey. Strateg. Change **26**(2), 193–212 (2017)
15. Lukkarinen, A., Teich, J.E., Wallenius, H., Wallenius, J.: Success drivers of online equity crowdfunding campaigns. Decis. Support Syst. **87**, 26–38 (2016)
16. Mamonov, S., Malaga, R.: Success factors in title III equity crowdfunding in the united states. Electron. Commer. Res. Appl. **27**, 65–73 (2018)
17. Mamonov, S., Malaga, R., Rosenblum, J.: An exploratory analysis of title ii equity crowdfunding success. Venture Cap. **19**(3), 239–256 (2017)
18. Mollick, E.: The dynamics of crowdfunding: an exploratory study. J. Bus. Ventur. **29**(1), 1–16 (2014)
19. Mollick, E., Nanda, R.: Wisdom or madness? comparing crowds with expert evaluation in funding the arts. Manage. Sci. **62**(6), 1533–1553 (2016)
20. Ordanini, A., Miceli, L., Pizzetti, M., Parasuraman, A.: Crowd-funding: transforming customers into investors through innovative service platforms. J. Serv. Manage. **22**(4), 443–470 (2011)
21. Ralcheva, A., Roosenboom, P.: On the road to success in equity crowdfunding. SSRN (2016)
22. Tuomi, K., Harrison, R.T.: A comparison of equity crowdfunding in four countries: implications for business angels. Strateg. Change **26**(6), 609–615 (2017)
23. Vismara, S.: Equity retention and social network theory in equity crowdfunding. Small Bus. Econ. **46**(4), 579–590 (2016)

How Content Features of Charity Crowdfunding Projects Attract Potential Donors?

Empirical Study of the Role of Project Images and Texts

DongIl Lee and JaeHong Park[(⊠)]

Kyung Hee University, HeogiDaero 24, Seoul, Korea
{godmay0610, jaehp}@khu.ac.kr

Abstract. This study investigates how the content features (e.g., images and texts) of donation projects affect potential donors' participation. We collect textual and visual contents from one of the largest online donation crowdfunding platforms in South Korea. To extract features from the content, we use Deep Learning models for images and Latent Dirichlet Allocation (LDA) topic modeling for text contexts. We then construct variables representing visual and textual features. Finally, we estimate the effects of our independent variables on donors' participation by using the Ordinary Least Squares (OLS) model. Our empirical results show that (1) Observing a small number of recipients in images attract more donors than a large number of recipients does; (2) Negative and positive emotions decrease potential donors' willingness to help compared to neutral emotion; (3) Positive emotion in the image moderates the number of recipients' negative effect; and (4) Since complex project description requires more effort to understand the recipients, potential donors are less likely to be engaged. Through this study, we hope to make contributions to the extant literature. In addition, our framework for content analysis will contribute to the future studies as we shed light on novel methodologies to measure image and text dimensions.

Keywords: Content analysis · Donation · Crowdfunding · Deep learning · Topic modeling

1 Introduction

Assume that two charity crowdfunding projects are displayed together under the same category—Kids—and that both represent a need for help funding surgery (Fig. 1). If you have $100 to spare for donations, which project would you choose to support?

The most salient features of the projects are the thumbnail images. A notable difference between the two is that we cannot see the recipient's face in left image, while we can clearly observe a face in right image. We can easily recognize that the former is more abstract than the latter. That is, we may feel stronger emotional engagement with donation project on the right. Therefore, which of the two do you think will receive more pledges?

© Springer Nature Switzerland AG 2019
J. J. Xu et al. (Eds.): WEB 2018, LNBIP 357, pp. 124–131, 2019.
https://doi.org/10.1007/978-3-030-22784-5_12

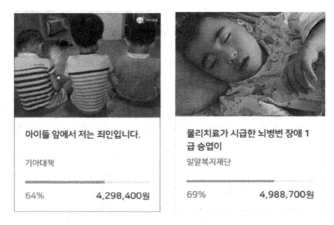

Fig. 1. Two donation crowdfunding projects

Recent studies on donation have indicated that empathy has a positive effect on charity (Eisenberg and Miller 1987), and this effect depends on how easy it is to envision another's situation (Paniculangara and He 2012). Also, potential donors experience more emotional reactions—empathy—to a focus on an identified single victim, while the identification of group has no effect on willingness to contribute (Jenni et al. 1997; Small et al. 2007; Kogut and Ritov 2011). Researchers named this tendency as "identified victim effect" and "singularity victim effect". Moreover, sad facial expression of recipients increases donation compared to happy expression. Likewise, emotion, number and identifiability of recipients in the content are important factor to attract more of the potential donors' participation. However, there is a lack of studies that found interaction effect between number of recipient and emotion on donors' participation. Furthermore, these studies have conducted their research under restricted experiment conditions and potentially do not directly measure the influence of project campaigns' content on the backers' contributions.

In our research context, not only visual content but textual content of charity crowdfunding projects also conveys important information for latent donor. According to cognitive and psychology theories of text comprehension (Goodman 1967; Gough 1972; Kintsch and van Dijk 1978), more complex textual content requires more effort to absorb and understand the context.

In this regard, we examine three research questions: (RQ1) How much does the number of recipients on the image influence donors' participation? (RQ2) How do recipients' emotions influence donors' participation? Then, (RQ2-1) Is there an interaction effect between number of recipient and emotion on donors' participation? (RQ3) How much does the text complexity of the project's description affect donors' participation?

To answer these questions, we collect textual and visual contents from one of the largest online donation crowdfunding platforms in South Korea. We analyze what kind of visual and textual contents appeal more to the potential donors. In order to extract the features from the content, we use Deep Learning models to detect human recipients and their emotion/gender classification from images and Latent Dirichlet Allocation (LDA) topic modeling to measure text complexity of text contexts. Then, we estimate their effects on donors' participation.

We thus hope to make both theoretical and practical contributions to the extant literature. Theoretically, we can contribute to the existing literature by measuring the effect of both visual and textual contents of donation projects on potential donor's participation. Specifically, we first investigate an effect of text complexity of the projects' description and an interaction effect between the number of recipient and their emotion. We also believe that this is the first study to examine how visual and textual contents of donation projects influence donors' participation (i.e., number of donors) using novel methodology with donor-generated data, while previous studies are usually based on experiment in laboratory setting. Furthermore, our framework for content analysis will contribute to the future studies as we shed light on novel methodologies using deep learning techniques and LDA topic modeling to measure image and text dimensions. Practically, this study will help crowdfunding project managers to raise donors' participation by placing emphasis on project content design.

2   Data

2.1   Data Source

We selected NAVER's Happybean as our research data source. Happybean is Korea's first online donation crowdfunding platform operated by NAVER's Happybean Foundation. On the Happybean platform, charity organizations are fully responsible for setting up a project and its contents. As a result, the quality of the projects' content may vary depending on who created them.

Our Happybean dataset consists of 663 projects. Since our research is based on humans, we exclude the projects under the Animal and Environment categories. All of the projects started and ended within the five-month period between July 2017 and December 2017.

2.2   Variables

Our main dependent variable is Donors, which represents the number of donors for each project. For our independent variables, we extract content features by using deep learning models and topic modeling. First, we use deep learning models for visual content. Deep learning approaches aim to automatically discover intricate structure in high-dimensional image data, making it possible to understand the semantic content of images. This has led to dramatically improved performance on image recognition and classification tasks. For detecting humans, we use the pre-trained object detection model, Single Shot MultiBox Detector (SSD) with inception-v2, provided by TensorFlow (Liu et al. 2016; Szegedy et al 2016; Ning et al. 2017). To classify emotion and gender, we use a well-trained classification model inspired by Xception networks (Chollet 2017; Arriaga et al. 2017). In Fig. 2, we provide the sample results of the above two models.

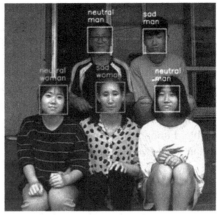

Fig. 2. Examples of human detection & emotion-gender recognition

We then construct variables representing visual features. *NumRecip* is the number of recipients on the project's thumbnail image. *EmoPos, EmoNeg,* and *EmoNeu* are dummy variables representing an image's overall emotion. Furthermore, *GenderRatio* (Number of woman/Total number of recipient) indicate gender ratio in the image. Moreover, we considered not only image features extracted through deep learning, but also other relevant image features. The first is *InnerNumImg,* which is the number of inner content images each project exhibits. Also, we noticed that the quality of the thumbnail images varies across the projects. Therefore, we also include *ImgQuality* to capture the quality of the images. We used the compressed JPG image file size as the *ImgQuality.*

For the text-related variable, *TextComplex* represents the complexity of the project description. To capture this, we computed topic models using the LDA implementation. There are two outputs from the LDA: (i) keyword sets for each topic and (ii) topic distribution for each document (i.e., project description). An important advantage of LDA is that researchers can read the automatically constructed keyword sets (i.e., topics) to understand the underlying topics. We use the latter output of the LDA model and compute its complexity using the Shannon entropy index. Let $p \in [0, 1]^d$ be the predicted confidence scores for a given text (i.e., d = 27). Then, the topic complexity is defined as:

$$\text{Textcomplex} = -\sum_i^d p_i \log(p_i)$$

Note that $\sum_i^d p_i = 1$ and the maximum value of complexity = $\log(d)$ when p is uniformly distributed. As the text content becomes focused on fewer elements, complexity gets smaller and eventually becomes 0 when $p_i = 1$ for some i.

As control variables, we include each project's duration (*Duration*), goal amount (*GoalAmt*) and dummy variables for categories (*BetterWorld, WorldShare, Human-right, Children, Eldery, SeasonalIssue*). Finally, we include the total pledged amount (TotDonAmt) of each project to control for the possible heterogeneity of project sizes.

Table 1 summarizes the variables used in the analysis and their descriptive statistics. We observe that the distributions of *Donors* are skewed; there are a few projects with a massive amount of participations, the majority of which received limited attention. Therefore, we use the log transformed value of *Donors* as the dependent variable. Furthermore, since some of our independent variables have different units of measurement, we also take their logs to level their scales.

Table 1. Descriptive statistics of dependent and independent variables.

Variables	Obs	Mean	Std. Dev.	Min	Max
Donors	663	237	334.903	2	2432
NumRecip	663	1.766	2.090	0	14
EmoPos	148	0.3514	0.479	0	1
EmoNeg	148	0.3716	0.485	0	1
TextComplex	663	0.6060	0.381	0	1
ImgQuality	663	524.51	1081.518	3.03	7478.35
InnerNumImg	663	3.811	2.583	0	15
GenderRatio	148	0.5063	0.448	0	1
Duration	663	59.39	24.779	19	143
GoalAmt	663	4533347	4338031	50000	20000000
TotDonAmt	663	240800000	473215000	14000	1873000000

* Some variables have 148 observations, since not all projects show the recipients' faces. Values are assigned to these variables only when the deep learning algorithm detects faces in the projects' thumbnail images.

3 Model and Empirical Results

In this study, we investigate how the content features (e.g., images and texts) of donation projects affect potential donors' participation. To do so, we measure the effects of our independent variables on the donors' participation by using the Ordinary Least Square (OLS) model. For the *i*-th project, our model can be written as:

$$
\begin{aligned}
Y_i(\ln(Donors_i)) \\
= \beta_0 + \beta_1 \ln(NumRecip_i) + \beta_2 EmoPos_i + \beta_3 EmoNeg_i \\
+ \beta_4 \ln(TextComplex_i) + \beta_5 EmoPos_i \times \ln(NumRecip_i) \\
+ \beta_6 EmoNeg_i \times \ln(NumRecip_i) + \beta_7 \ln(ImgQuality_i) \\
+ \beta_8 \ln(InnerNumImg_i) + \beta_9 GenderRatio_i + \beta_{10} \ln(Duration_i) \\
+ \beta_{11} \ln(GoalAmt_i) + \beta_{12} \ln(TotDonAmt_i) + \beta_{13} Categoty_i + \varepsilon_i
\end{aligned}
$$

3.1   Main Results

Table 2 summarizes the main empirical results from our research model. Model 1 shows the results with the control variables only. Since not all project images show the recipients' faces, some variables (e.g. *EmoPos*) have 148 observations. So, we preferentially put three image variables, *ImgQuality, NumRecip, InnerNumImg,* and *TextComplex,* in Model 2. Finally, we put all variables and interaction term in Model 3.

Table 2. Results for three models

Variables	Model 1		Model 2		Model 3	
	Coef	S.E	Coef	S.E	Coef	S.E
ln(NumRecip)			-0.1968^{***}	0.0672	-0.7327^{**}	0.3220
EmoPos					-1.1876^{**}	0.5370
EmoNeg					-0.8242	0.5089
ln(TextComplex)			-0.7287^{***}	0.1605	-0.6752^{*}	0.3451
EmoPos × ln (NumRecip)					0.7516^{*}	0.4123
EmoNeg × ln (NumRecip)					0.4389	0.3878
ln(ImgQuality)			0.0627^{**}	0.0256	0.1083^{*}	0.0570
ln(InnerNumImg)			0.1908^{***}	0.0682	0.2806^{*}	0.1513
GenderRatio					0.2359	0.1955
ln(Duration)	0.3585^{***}	0.1039	0.2904^{***}	0.1017	0.1826	0.2371
ln(GoalAmt)	0.4046^{***}	0.0415	0.4013^{***}	0.0416	0.2782^{***}	0.0934
ln(TotDonAmt)	0.2027^{***}	0.0237	0.1685^{***}	0.0242	0.2151^{***}	0.0512
Category	YES	YES	YES	YES	YES	YES
Constant	-6.3953^{***}	0.5993	-5.4929^{***}	0.6546	-5.0027^{***}	0.0096
Adjusted R-squared	0.3608		0.4601		0.4705	
Number of project	663		663		148	

*** p < 0.01, ** p < 0.05, * p < 0.1

We first considered the visual features, the quality of the image (*ImgQuality*) has a significant and positive effect. Increasing the image quality by 10% increases the number of donors by 1.08% in Model 3. The number of the project's inner images (*InnerNumImg*) also has a significant and positive effect. In Model 3, increasing the number of the project's inner images by 10% leads to a 2.80% increase in the number of donors. On the other hand, the number of recipients in the image has a significant but negative effect. Having 10% more recipients in an image (*NumRecip*) decreases the number of donors by 7.32% based on neutral emotion. This finding indicates that observing a single recipient in the images provokes a stronger willingness to help recipient than a large number of recipients in the image (Kogut and Ritov 2011). We also find that the dummy variables for positive emotion (*EmoPos*), have a significant and negative effect based on the neutral emotion in Model 3. If an image has positive emotion, the number of donors decreased by 6.95%. These results indicate that negative emotions may depress potential donors and thus lead to little participation.

Finally, we found that interaction effect between the number of recipients and positive emotion in the image has a significant and positive effect. If an image shows positive emotion and have 10% more recipients, the number of donors increased by 0.19%. This finding indicates that positive emotion in the image makes *NumRecip*'s negative effect into positive effect.

We observed that the text complexity (*TextComplex*) of a project description has significantly negative effects on donors' participation. Increasing 10% of the text complexity can decrease the number of donors by 6.75% in Model 3. One explanation is that, since the complex project description requires more effort to understand the recipients, potential donors are less likely to be engaged.

4 Future Plan

In this research, we empirically analyzed 663 donation projects from 317 charities to estimate the effects of visual and textual contents on donors' participation and the number of donors. We used machine learning techniques—deep learning and topic modeling—to construct features from unstructured visual and textual content data.

However, this study also has limitations. First, the textual content is relatively less examined than visual content in this study. So, we will extract another text complexity feature. In the future study, we may use word2vec by maximizing the predicted probability of words co-occurring within a small window of consecutive words (e.g., five words before and after the focal word). Using the trained word2vec model, we compute the probability of each sentence in a given project based on the order of words, and the mean value of those probabilities will be used as another topic complexity feature. Furthermore, we will analyze the sentiment of the project descriptions to examine that sentiment on text and emotion on image have similar effect on donor's participation. Second, we will also develop theoretical arguments to explain and support our findings. We believe that such a theoretical argument will make our paper more robust.

References

Arriaga, O., Valdenegro-Toro, M., Plöger, P.: Real-time convolutional neural networks for emotion and gender classification. arXiv preprint arXiv:1710.07557 (2017)

Basil, D.Z., Ridgway, N.M., Basil, M.D.: Guilt and giving: a process model of empathy and efficacy. Psychol. Mark. **25**(1), 1–23 (2008)

Chollet, F.: Xception: deep learning with depthwise separable convolutions. arXiv preprint arXiv:1610.02357 (2017)

Eisenberg, N., Miller, P.A.: The relation of empathy to prosocial and related behaviors. Psychol. Bull. **101**(1), 91–119 (1987)

Goodman, K.S.: Reading: a psycholinguistic guessing game. J. Read. Spec. **6**(4), 126–135 (1967)

Gough, P.B.: One second of reading. Visible Lang. **6**(4), 291–320 (1972)

Jenni, K.E., Loewenstein, G.: Explaining the "identifiable victim effect". J. Risk Uncertain. **14**(3), 235–257 (1997)

Paniculangara, J., He, X.: Empathy, donation, and the moderating role of psychological distance. Adv. Consum. Res. **40**, 250–254 (2012)

Kintsch, W., van Dijk, T.A.: Toward a model of text comprehension and production. Psychol. Rev. **85**(5), 363–394 (1978)

Kogut, T., Ritov, I.: The identifiable victim effect: causes and boundary conditions. In: Oppenheimer, D.M., Olivola, C.Y. (eds.) The Science of Giving: Experimental Approaches to the Study of Charity, pp. 133–145. Psychology Press, New York (2011)

Liberman, N., Trope, Y.: The role of feasibility and desirability considerations in near and distant future decisions: a test of temporal construal theory. J. Pers. Soc. Psychol. **75**(1), 5–18 (1998)

Liu, W., et al.: SSD: single shot multibox detector. In: Leibe, B., Matas, J., Sebe, N., Welling, M. (eds.) ECCV 2016. LNCS, vol. 9905, pp. 21–37. Springer, Cham (2016). https://doi.org/10.1007/978-3-319-46448-0_2

Loewenstein, G., Small, D.A.: The scarecrow and the tin man: the vicissitudes of human sympathy and caring. Rev. Gen. Psychol. **11**(2), 112–126 (2007)

Ning, C., Zhou, H., Song, Y., Tang, J.: Inception single shot multibox detector for object detection. In: 2017 IEEE International Conference on Multimedia & Expo Workshops (ICMEW), pp. 549–554. IEEE, July 2017

Simandan, D.: Proximity, subjectivity, and space: rethinking distance in human geography. Geoforum **75**, 249–252 (2016)

Small, D.A., Loewenstein, G., Slovic, P.: Sympathy and callousness: the impact of deliberative thought on donations to identifiable and statistical victims. Organ. Behav. Hum. Decis. Process **102**(2), 143–153 (2007)

Szegedy, C., Vanhoucke, V., Ioffe, S., Shlens, J., Wojna, Z.: Rethinking the inception architecture for computer vision. In: Proceedings of the IEEE Conference on Computer Vision and Pattern Recognition, pp. 2818–2826 (2016)

Do Achievement Goals and Work Nature Affect Contributor Performance in Gamified Crowdsourcing? An Exploratory Study in an Academic Setting

Philip Tin Yun Lee[1（✉）], Richard Wing Cheung Lui[2],
and Michael Chau[1]

[1] The University of Hong Kong, Pok Fu Lam, Hong Kong
phil0127@hku.hk, mchau@business.hku.hk
[2] The Hong Kong Polytechnic University, Kowloon, Hong Kong
cswclui@comp.polyu.edu.hk

Abstract. Many studies have demonstrated the benefits of gamification in the context of crowdsourcing. However, not every user benefits equally from gamification. Most of the current studies focused on the game elements of gamified systems. Scant attention has been paid to the factors related to the users (Koivisto and Hamari 2014; Morschheuser et al. 2016). University students will be recruited to trial a gamified crowdsourcing system for two weeks. Our study aims to explore whether achievement goal orientations influence user performance in gamified crowdsourcing systems. In addition, certain types of crowdsourcing require creativity, whereas tasks of other types of crowdsourcing can be done mechanically. The achievement goals may also affect user performance in different tasks. Our study explores whether users' achievement goals affect their performance in homogeneous and heterogeneous tasks respectively in the context of gamified crowdsourcing. Results of our study will contribute to the expanding literature on whether gamification works on all people. The results will also help us understand more about the behavior of users with different achievement goals in gamified crowdsourcing systems.

Keywords: Crowdsourcing · Gamification · Achievement goals · Creativity · Brainstorming

1 Introduction

Crowdsourcing harnesses the intelligence and efforts of the crowd. Crowdsourcing systems exist in various forms. For example, organizations can outsource trivial tasks to the crowd through crowdsourcing systems like Amazon Mechanical Turk. Learners can post questions, answer enquiries and rate answers on crowdsourcing websites such as Stack Overflow and ResearchGate. Travelers can also rate hotels and tourist spots on crowdsourcing platforms, e.g. TripAdvisor. The underlying principle of crowdsourcing is that contributors can follow their own preferences and choose their own tasks freely (Geiger and Schader 2014).

© Springer Nature Switzerland AG 2019
J. J. Xu et al. (Eds.): WEB 2018, LNBIP 357, pp. 132–140, 2019.
https://doi.org/10.1007/978-3-030-22784-5_13

Crowdsourcing applications can be generally classified into four types, including crowd solving, crowd creation, crowd processing and crowd rating (Geiger et al. 2012). They differentiate from each other in two dimensions, i.e. values derived from contributions and values differentiated among contributions (see Fig. 1). Contributions of crowd processing and crowd rating are valued equally. Each of the contribution leads to identical or similar rewards. The tasks of crowd processing and crowd rating are usually rather homogeneous. On the other hand, values of each contribution in crowd solving and crowd creation are not the same. Tasks of crowd solving and crowd creation are heterogeneous. They usually demand certain extents of contributors' creativity and innovation. Contributors who provide quality work receive better rewards.

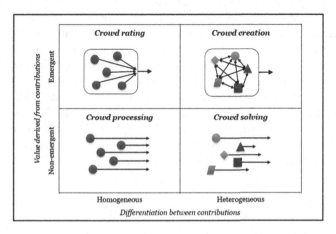

Fig. 1. Four types of crowdsourcing information systems (Adapted from Geiger et al. (2012)).

Values derived from contributions can be classified as emergent and non-emergent. Crowd rating and crowd creation are considered as emergent crowdsourcing systems where fusion of contributions constitutes a better output. Crowd processing and crowd solving are non-emergent. Integration of their contributions brings no extra values.

The contributors of most crowdsourcing systems are volunteers. Thus, how to motivate the contributors is an important question for system designers as well as researchers. Gamification has recently been adopted with the aim of motivating users of crowdsourcing systems.

Various studies have demonstrated the benefits of gamification in the context of crowdsourcing. Some examples include an increase of engagement (Itoko et al. 2014; Vasilescu et al. 2014) and an improvement in work quality (Eickhoff et al. 2012; Goncalves et al. 2014). However, most of the studies attended to game elements of systems. Scant attention was paid to the factors related to the users (Koivisto and Hamari 2014; Morschheuser et al. 2016). According to Morschheuser et al. (2016)'s review paper, only one paper explored the difference in behaviors between different user groups: Itoko et al. (2014) investigated the effectiveness of game affordances for young and old users in a gamified proofreading system. Our study sheds light on the difference in effectiveness of gamification for people with different achievement goals.

Furthermore, different types of crowdsourcing require different competence. Whereas crowd processing and crowd rating tasks can be completed rather mechanically, crowd solving and crowd creating tasks require contributors' creativity. Performance in different types of tasks should be evaluated respectively.

An exploratory study will be conducted among university students who are enrolled in the same course. They will trial a gamified crowdsourcing system for two weeks as an exercise of the course. Their achievement goals (Elliot and McGregor 2001) and creative self-efficacy (Tierney and Farmer 2002) will be measured before the trial, and their performance in both homogeneous and heterogeneous tasks over the two-week duration will be recorded. The study aims to explore whether achievement goal orientations influence user performance in gamified crowdsourcing applications. We propose that a point system and a leaderboard, two common motivational affordances in games, create an environment through which users can compare themselves with others and gain senses of achievement. The users with a stronger performance-approach goal will perform better, since they are more easily influenced by the positive senses of achievement gained from social comparison. On the other hand, the users with a strong mastery-avoidance goal tend to avoid making mistakes. We propose that this behavior may limit their performance in heterogeneous tasks. Results of our study will contribute to the expanding literature on whether gamification works on all people. The results will help us understand more about the behavior of users with different achievement goals in gamified crowdsourcing systems.

2 Literature Review

2.1 Gamification

Gamification is defined as using the game elements in non-game contexts (Deterding et al. 2011). Gamification has been applied in different areas such as commerce, education and health (For a detailed review, see Hamari et al. (2014)). The most commonly implemented features of gamification in crowdsourcing context were point systems and leaderboards (Morschheuser et al. 2016). These two motivational affordances help create a gaming environment and facilitate competition among participants.

2.2 Individual Difference Towards Gamification

Not everyone perceives and benefits equally from gamification. A number of researchers have shown that users' attitudes towards gamified systems differ. Montola et al. (2009) interviewed contributors of a photo sharing mobile application, and identified three different types of attitudes. They categorized the contributors as indifferent users, confused users and appreciative users. Eickhoff et al. (2012) suggested that contributors can be classified as either entertainment-motivated or money-motivated. Gamified crowdsourcing systems should be customized for contributors with different motivations. Hamari (2013) found that earning badges did not significantly affect all users' behaviors in a peer-to-peer trading service system. Only users who kept an eye on their own badges and compared them against other users were positively influenced by the badges. Koivisto and Hamari (2014) discovered that female users perceived a gamified exercise-tracking system more playful than men did. The female benefited more from the system in social influence, reciprocal benefits and recognition.

2.3   Achievement Goal

An achievement goal is defined as "an integrated pattern of beliefs, attributions, and affect that produces the intentions of behavior" (Ames 1992, p. 261). It is concerned with the underlying aims of achievement behavior. There are two major kinds of achievement goals of students, namely mastery goals and performance goals. These goals represent different concepts of success and approaches adopted to reach the success (Ames 1992). The mastery goal is concerned with people improving their abilities and mastering new skills. On the other hand, the performance goal focuses on that individuals gain a sense of achievement through comparison among their counterparts (Dweck 1986).

More recent papers incorporated the concepts of approach-avoidance motivation into the previous two achievement goals (Elliot and McGregor 2001). Approach motivation features active acquisition of positive outcomes, whereas avoidance motivation highlights avoidance of negative possibilities. Thus, this framework outlined four achievement motivations, namely mastery-approach, mastery-avoidance, performance-approach, and performance-avoidance. People differ in their achievement motivations.

2.4   Creativity

Creativity plays an important role in the context of crowdsourcing. Particularly, it is required by tasks of crowd creation and crowd solving. Individual creativity can be indirectly measured by creative self-efficacy, given that people who are more creative possess stronger creative self-efficacy (Tierney and Farmer 2002). Several studies revealed possible linkages between creativity and achievement goals. Gong et al. (2009) empirically showed that creativity and job performance were positively correlated. They also suggested that mastery-approach goal enhanced creativity over time. Hirst et al. (2009) suggested that team learning behavior was a moderator of the relationship between a mastery-approach goal and creativity. Huang and Luthans (2015) indicated an indirect effect of learning goals on creativity when people think and behave independently. These studies show a strong relationship between achievements goals and creative self-efficacy in various environments. These studies reveal possible effects of achievement goals to performance in heterogeneous tasks in gamified settings.

3 Hypothesis Development

Gamification incorporates game affordances into a non-game context. The gaming environmental cue of a gamified system may serve as a stimulus of achievement to users. If the users possess a strong achievement goal, the cognitive link between achievement stimuli and the achievement goal is closer (Bargh 1990; McClelland et al. 1953). In the presence of the achievement stimuli, the users will spend more efforts on the tasks in a reflexive manner (Bargh et al. 2001; Shah 2003).

Point systems and leaderboards cultivate a competing environment where users can interact and compare with other users. The motivational affordances render a larger exposure of the users to competition. They also help promote positive judgments among users. Thus, the competition forms an achievement stimulus that is more associated with the performance-approach goal, since the performance-approach goal features positive possibilities of social comparison with counterparts. Also, given that the different nature of homogeneous and heterogeneous tasks may affect user performance. Performance in these tasks should be considered respectively. We therefore hypothesize that:

H1a: If users have a stronger performance-approach goal, then they will perform better in heterogeneous tasks in a gamified crowdsourcing environment.

H1b: If users have a stronger performance-approach goal, then they will perform better in homogeneous tasks in a gamified crowdsourcing environment.

On the other hand, these motivational affordances may bring less advantage to people with a strong mastery goal, since these people place more emphasis on personal development. Furthermore, people with a strong mastery-avoidance goal tend to prevent themselves from negative judgment of their abilities. They aim at making no mistakes. This behavior may limit their performance in creativity work, given that criteria for correct answers in heterogeneous work is less objective and explicit. Such behavior, however, may not influence their performance in homogenous tasks. Hence, we hypothesize that:

H2a: If users have a stronger mastery-avoidance goal, then they will perform worse in heterogeneous tasks in a gamified crowdsourcing environment.

H2b. Even if users have a stronger mastery-avoidance goal, they will not necessarily perform worse in homogeneous tasks in a gamified crowdsourcing environment.

4 Methods and Data Analysis

4.1 The Gamified Crowdsourcing System

A gamified crowdsourcing system developed by an IT company in Hong Kong will be used in our exploratory study. Two motivational affordances – a point system and a leaderboard – are implemented in the system. Users can choose to work on any outstanding tasks (see Fig. 2). Heterogeneous tasks refer to solutions to the brainstorming tasks in the system. These brainstorming tasks are concerned with daily conversation in different business contexts. Questions in the tasks are designed in a way that little specialized knowledge is required to complete them. The questions are comparable to

Malaga (2000)'s question for brainstorming sessions: "Produce a list of as many new delicious ice cream flavors as possible" (p. 132). The answers will be reviewed by other users (see Fig. 3). Answers that are reviewed and endorsed by more than 3 users are considered as correct answers. The review tasks are considered as homogeneous tasks. The users receive points for correct answers, and additional points will be given for unseen answers detected by the system. In addition, not every task has the same rewards. Some more challenging tasks offer more points to contributors.

The system has not been publicly launched, so the system is new to all users. The outstanding tasks in the system are brainstorming and review assignments that are related to daily conversation in different business application domains.

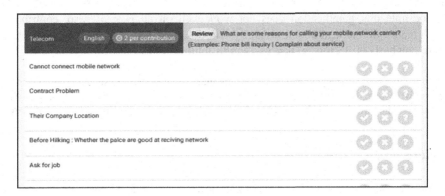

Fig. 2. An example of a brainstorming task in the gamified crowdsourcing system.

Fig. 3. An example of a review task in the gamified crowdsourcing system.

4.2   Data Collection and Analysis

Undergraduate students enrolled in a computer science course will be recruited to participant the study as an exercise to understand system design of crowdsourcing systems. They will trial the gamified crowdsourcing system for two weeks. They will be told that the crowdsourcing work in the system will be used to support a chatbot development project. Before the trial, they will fill in online questionnaires which measure their achievement goals and creative self-efficacy. Questions of achievement goals are adopted from Elliot and McGregor (2001), whereas questions of creative self-efficacy are adopted from Tierney and Farmer (2002). The scores of brainstorming tasks and review tasks are recorded respectively. These scores are dependent variables.

To test the hypotheses, we follow Elliot and McGregor (2001) to conduct a simultaneous multiple regression analysis to predict each score from the four achievement goal orientations. The creative self-efficacy will be controlled in each analysis.

4.3 Limitations

The study involves some limitations. First, the participants are enrolled in a particular course. The variety of population may be limited. Also, it is unclear whether the two-week trial is long enough. Furthermore, our study measures achievement goals for learning. The choice of measurement is considered to be appropriate, given that the trial of the system is an exercise of the course. The measurement, however, should be adapted for study in other fields.

5   Ongoing Work

Participants with strong performance-approach goals are expected to perform better in all tasks. This will indicate that people with a stronger performance-approach goal benefit more from a gamified setting. Participants with a strong mastery-avoidance goal are expected to perform worse in brainstorming tasks. These possible results will demonstrate that gamification does not work well on all people. Also, contributor performance differs in different tasks. System designers and researchers should place more emphasis on factors related to users in future studies.

Acknowledgement. The gamified system in this study is developed and supported by Imago AI (https://www.imago.ai/). We would like to thank for their support to our research project. The research is supported in part by the HKU Seed Funding for Basic Research (No. 104004856).

References

Ames, C.: Classrooms: goals, structures, and student motivation. J. Educ. Psychol. **84**(3), 261–271 (1992)

Bargh, J.A.: Auto-motives: preconscious determinants of thought and behavior. In: Higgins E.T., Sorrentino R.M. (eds.) Handbook of motivation and cognition. Hillsdale, pp. 1–40 (1990)

Bargh, J.A., Gollwitzer, P.M., Lee-Chai, A., Barndollar, K., Troetschel, R.: The automated will: nonconscious activation and pursuit of behavioral goals. J. Pers. Soc. Psychol. **81**, 1014–1027 (2001)

Deterding, S., Dixon, D., Khaled, R., Nacke, L.: From game design elements to gamefulness: defining gamification. In: Proceedings of the 15th International Academic MindTrek Conference: Envisioning Future Media Environments, pp. 9–15. ACM, September 2011

Dweck, C.S.: Motivational processes affecting learning. Am. psychol. **41**(10), 1040–1048 (1986)

Eickhoff, C., Harris, C.G., de Vries, A.P., Srinivasan, P.: Quality through flow and immersion: gamifying crowdsourced relevance assessments. In: Proceedings of the 35th International ACM SIGIR Conference on Research and Development in Information Retrieval, pp. 871–880. ACM, August 2012

Elliot, A.J., McGregor, H.A.: A 2×2 achievement goal framework. J. Pers. Soc. Psychol. **80**(3), 501–519 (2001)

Geiger, D., Rosemann, M., Fielt, E., Schader, M.: Crowdsourcing information systems-definition, typology, and design. In: Proceedings of the 33rd International Conference Information System, pp. 1–11 (2012)

Geiger, D., Schader, M.: Personalized task recommendation in crowdsourcing information systems - Current state of the art. Decis. Support Syst. **65**, 3–16 (2014)

Goncalves, J., Hosio, S., Ferreira, D., Kostakos, V.: Game of words: tagging places through crowdsourcing on public displays. In: Proceedings of the 2014 conference on Designing interactive systems, pp. 705–714. ACM, June 2014

Gong, Y., Huang, J.C., Farh, J.L.: Employee learning orientation, transformational leadership, and employee creativity: the mediating role of employee creative self-efficacy. Acad. Manag. J. **52**(4), 765–778 (2009)

Hamari, J.: Transforming homo economicus into homo ludens: a field experiment on gamification in a utilitarian peer-to-peer trading service. Electron. Commer. Res. Appl. **12**(4), 236–245 (2013)

Hamari, J., Koivisto, J., Sarsa, H.: Does gamification work?–a literature review of empirical studies on gamification. In: Proceedings of 2014 47th Hawaii international conference on system sciences (HICSS), pp. 3025–3034. IEEE, January 2014

Hirst, G., Van Knippenberg, D., Zhou, J.: A cross-level perspective on employee creativity: goal orientation, team learning behavior, and individual creativity. Acad. Manag. J. **52**(2), 280–293 (2009)

Huang, L., Luthans, F.: Toward better understanding of the learning goal orientation–creativity relationship: the role of positive psychological capital. Appl. Psychol. **64**(2), 444–472 (2015)

Itoko, T., Arita, S., Kobayashi, M., Takagi, H.: Involving senior workers in crowdsourced proofreading. In: Stephanidis, C., Antona, M. (eds.) UAHCI 2014. LNCS, vol. 8515, pp. 106–117. Springer, Cham (2014). https://doi.org/10.1007/978-3-319-07446-7_11

Koivisto, J., Hamari, J.: Demographic differences in perceived benefits from gamification. Comput. Hum. Behav. **35**, 179–188 (2014)

McClelland, D.C., Atkinson, J.W., Clark, R.W., Lowell, E.L.: The Achievement Motive. Appleton-Century-Crofts, New York (1953)

Malaga, R.A.: The effect of stimulus modes and associative distance in individual creativity support systems. Decis. Support Syst. **29**(2), 125–141 (2000)

Montola, M., Nummenmaa, T., Lucero, A., Boberg, M., Korhonen, H.: Applying game achievement systems to enhance user experience in a photo sharing service. In: Proceedings of the 13th International MindTrek Conference: Everyday Life in the Ubiquitous Era, pp. 94–97. ACM, September 2009

Morschheuser, B., Hamari, J., Koivisto, J.: Gamification in crowdsourcing: a review. In: Proceedings of 2016 49th Hawaii International Conference on System Sciences (HICSS), pp. 4375–4384. IEEE, January 2016

Shah, J.Y.: Automatic for the people: How representations of significant others implicitly affect goal pursuit. J. Pers. Soc. Psychol. **84**, 661–681 (2003)

Tierney, P., Farmer, S.M.: Creative self-efficacy: Its potential antecedents and relationship to creative performance. Acad. Manag. J. **45**(6), 1137–1148 (2002)

Vasilescu, B., Serebrenik, A., Devanbu, P., Filkov, V.: How social Q&A sites are changing knowledge sharing in open source software communities. In: Proceedings of the 17th ACM Conference on Computer Supported Cooperative Work & Social Computing, pp. 342–354. ACM, February 2014

Is a Blockchain-Based Game a Game for Fun, or Is It a Tool for Speculation? An Empirical Analysis of Player Behavior in Crypokitties

Jaehwan Lee[1], Byungjoon Yoo[1], and Moonkyoung Jang[2(✉)]

[1] Seoul National University, Seoul 08826, South Korea
[2] Korea University, Seoul 02841, South Korea
jmoonk25@korea.ac.kr

Abstract. The market growth and popularity of Blockchain technology in various fields has been astonishing. The gaming industry is one of the prominent industries that applies Blockchain technology. However, there is some concern about the speculative aspect of Blockchain-based games. Therefore, this paper examines the effects of speculative and enjoyable aspects on users' playing behavior in Blockchain-based games. For this, we developed a web crawler and collected data from Cryptokitties, one of very first successful Blockchain-based games. To analyze our dataset, we conducted fixed panel regression. The results indicate that the external aspects of the game are negatively related to item selling, whereas the internal aspects of the game are positively related to item selling in the Blockchain-based game. This result suggests that playing Blockchain-based games is a mix of behavior with enjoyable and speculative aspects. As such, it remains unclear whether users play Blockchain-based games just for fun or as a form of speculation.

Keywords: Blockchain · Blockchain-based game · Cryptokitties

1 Introduction

Cryptocurrency has become more than a digital asset. In the early stages of cryptocurrency, people handled it as a kind of digital currency that could only be use via the Internet. However, as people have come to understand its unique characteristics, which allows for a decentralized payment system, it started to attract massive public attention. There are now more than 1,600 types of cryptocurrency including Bitcoin, Ethereum, EOS, etc. [6]. Cryptocurrency is also garnering academic interests and scholars have examined its viability as currency [8], its speculative characteristics [1, 4], determinants or trend of cryptocurrency price [2, 11, 15].

Blockchain technology is one of the key cryptocurrency technologies that allows the decentralized payment system [14]. The market size of Blockchain technology is estimated to be about 1 trillion in valuation and growing [3]. Blockchain technology is not used just for cryptocurrency, but for diverse applications such as trading, ecosystem, security, social service [19]. In various fields, the gaming industry is prominent. There are now more than 300 Blockchain-based games and more continue to be

© Springer Nature Switzerland AG 2019
J. J. Xu et al. (Eds.): WEB 2018, LNBIP 357, pp. 141–148, 2019.
https://doi.org/10.1007/978-3-030-22784-5_14

launched [7]. Users of Blockchain-based games can spend and earn cryptocurrency during their gameplay.

As the Blockchain-based game industry continues to grow, people are becoming concerned with the speculative or addictive features of Blockchain-based games [9]. These kinds of worries existed at the beginning of cryptocurrency and many researchers have investigated different aspects of cryptocurrency. However, there is a lack of academic research on the enjoyable or speculative aspects of Blockchain-based games. Therefore, this paper investigates the effect of enjoyable and speculative aspects on users' playing behavior in Blockchain-based games. To do this, we developed a web crawler to gather users' transactions in their open ledger related to Cryptokitties (www. cryptokitties.co), an early Blockchain-based game that was released in March 2017 and was one of the very first Blockchain-based games to experience success [10].

The remainder of this paper is organized as follows. In Sect. 2, a research model and relevant hypotheses are suggested. Section 3 explains the data gathered from Cryptokitties and the analyzing method used to consider the features of our data. Section 4 shares and discusses preliminary results. Finally, Sect. 5 presents the conclusions of this research and directions for future research.

2 Research Hypothesis

Blockchain-based games are different from traditional games in several aspects. The main difference is the interoperability across games. Traditional games allow their assets to be used only in their own game, but the assets of Blockchain-based games are interoperable across Blockchain-based games. Unlike traditional games, Blockchain technology allows high asset security and open trading among users even across games. It is possible for users to profit because users can earn additional cryptocurrency by getting or receiving game items in Blockchain-based games.

In addition, the price of game items can go up in price regardless of cryptocurrency price, so users can earn extra cryptocurrency in the game. Therefore, users who have speculative purpose are likely to play games to gain extra profit by converting game items into cryptocurrency. These users pay attention to the price and the total volume of cryptocurrency for buying or selling game items. Speculative users try to achieve the maximum benefits from market fluctuations, so when the price is high and the trade volume is huge, they are willing to wait for potential bigger benefits. Therefore, the hypotheses are as follows.

Hypothesis 1: The price of cryptocurrency is negatively related to the number of items selling in Blockchain-based games.

Hypothesis 2: The total volume of cryptocurrency is negatively related to the number of items selling in Blockchain-based games.

In addition, to distinguish speculative and enjoyable behaviors in the blockchain-based game, we assume the behaviors related to the game itself are related to enjoyable behaviors such as in-game purchases, special functions in the game, interactions between users, etc. These in-game behaviors make the game enjoyable for users and keeps them playing. Selling in-game items also relates to in-game aspects, so the internal aspect of games will be related to item selling. The frequency or amount of item purchase presents the degree of immersion [12, 13, 17]. Highly immersed users will spontaneously play the game more using in-game functions. Accordingly, our hypothesis is as follows.

Hypothesis 3: Item purchase is positively related to item selling in Blockchain-based games.

Furthermore, users who use unique functions of the game will be more likely to feel satisfaction [16]. Users who are satisfied with unique features of the game will be more likely to use in-game functions. Thus, the hypothesis is formulated as follows:

Hypothesis 4: Unique game functions are positively related to item selling in Blockchain-based games.

As people enjoy the game for its social-interaction activities, many games set several items to encourage interaction with other players [18]. Accordingly, social-interaction playing behaviors can be described as playing behaviors in which users enjoy the game through interactions with others. Therefore, the relationship between social-interaction playing behaviors and item-selling behavior is investigated as follows.

Hypothesis 5: Social interactions are positively related to item selling in Blockchain-based games.

3   Research Method

3.1   Blockchain-Based Game

The dataset used in this paper is user-level transaction log data collected from the Cryptokitties website (www.cryptokitties.co). Cryptokitties is one of the very first blockchain-based games; it was released in March 2017. Using Ethereum blockchain, users are able to buy, collect, breed, present, and sell digital cats with unique characteristics made by a combination of certain attributes, or "cattributes." Figure 1 shows the main page of Cryptokitties.

Fig. 1. Main page of Cryptokitties

There are 4 billion possible variations of Cryptokitties, and a new Cryptokitty was released every 15 min until November 2018. Because all Cryptokitties are linked to the Ethereum, each new cat is assigned a generation. A limited number of highly desirable early - generation kitties had the highest sales prices. The total trading amount for Cryptokitty purchases was about 6.7 billion dollars until December 2017. Due to the rocketing number of Cryptokitties users, the Ethereum network experienced delays in 2018 [5]. Figure 2 presents screenshots of an individual user's web page and a kitty in Cryptokitties. Users can see their own kitties and their cattributes in the individual kitty page.

Fig. 2.   User page and kitty page in Cryptokitties

3.2   Variables

We developed a web crawler using Python to collect all transactional records from 1 December 2017 till 31 July 2018. The raw dataset contains all Ethereum transaction behaviors of 60,000 users. Among these datasets, we selected the data related to Cryptokitties transactions and analyzed these data, which numbered more than 100,000 rows. The key variables are as follows (Table 1).

Table 1.   Summary of key variables

Variable			Definition
Dependent variable	$SellKitties_{i,t}$		The number of Cryptokitties that a player i sells at time t
Independent variable	External	$EtherPrice_t$	The price of Ethereum at time t
		$EtherVlm_t$	The total volume of Ethereum at time t
	Internal	$Purchase_{i,t}$	The number of Cryptokitties that a player i buys at time t *Immersion*
		$Siring_{i,t}$	The Number of Cryptokitties that a player i sires at time t *Unique game function*
		$Gifting_{i,t}$	The Number of Cryptokitties that a player i presents at time t, *Social interaction*

To investigate the effects of enjoyable and speculative aspects on users' playing behavior in Cryptokitties, we considered the effect of external and internal factors. We assumed that users who consider speculative features of Cryptokitties as important are likely to pay attention to the price or the total volume of Ethereum. The price and total volume of Ethereum are considered external aspects. Figure 3 depicts the price trend of Ethereum within our data time window. The Ethereum price fluctuated dramatically.

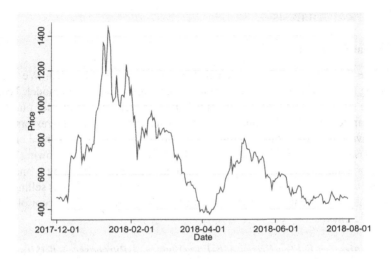

Fig. 3. Price of Ethereum

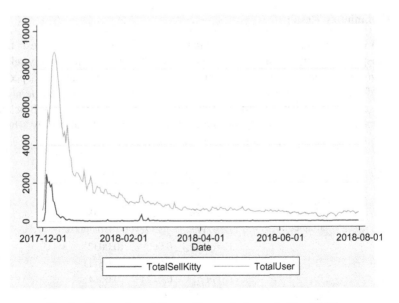

Fig. 4. The number of item selling and players in Cryptokitties

To investigate the effect of enjoyable aspects of Cryptokitties, we assumed that users who consider the game enjoyable are likely to use game functions often to play. Therefore, the game functions are considered internal aspects—the numbers of Cryptokitties a player buys, sires, and presents. Figure 4 illustrates the number of users and Cryptokitties that users sold. The numbers soared around the end of December 2017 and plummeted soon after.

4 Empirical Analysis

4.1 Research Model

To estimate the suggested factors, panel data models were conducted. We formally tested using techniques of the fixed effect model and the random effect model. To check whether the random effect model or the fixed effect model was suitable for our dataset, the Hausman test was carried out and its result indicated that the regressors are not correlated with the error terms. Therefore, our estimation incorporates fixed effects for users to consider unobserved characteristics of individual users. The following equation captures our econometric model. In this equation, users are indexed by i, and time is indexed by t. β_i is coefficients' estimates for the number of Cryptokitties selling at time t. μ_i is for the individual cross-sectional effect, and the error term $\varepsilon_{i,t}$ controls for the idiosyncratic effects.

$$SellKitties_{i,t} = \beta_1 EtherPrice_{i,t} + \beta_2 EtherVlm_{i,t} + \beta_3 Purchase_{i,t} + \beta_4 tSiring_{i,t} \\ + \beta_5 Gifting_{i,t} + \mu_i + \varepsilon_{i,t}$$

4.2 Preliminary Result

The result of our main model is demonstrated in Table 2. Our finding suggests that the external aspects (the price and the total volume of Ethereum) are negatively related to item selling (Hypotheses 1 and 2 are supported). In addition, the internal aspects (the number of item purchase, siring, and gifting) are positively related to item selling (Hypotheses 3, 4, and 5 are supported). The results indicate that users play Blockchain-based games for fun as well as speculation. Therefore, it is difficult to say whether a Blockchain-based game is just a game for fun or a tool for speculation.

Table 2. Main results from fixed effect analysis

Dependent variable		$SellKitties_{i,t}$
Independent variables	$EtherPrice_t$	-0.0006^{***} (0.0000)
	$EtherVlm_t$	$-3.12e{-}11^{***}$ (0.0000)
	$Purchase_{i,t}$	0.0669^{***} (0.0032)
	$Siring_{i,t}$	2.5421^{***} (0.0811)
	$Gifting_{i,t}$	0.0124^{***} (0.0014)
Observation		301,591
Number of users		34,723

5   Conclusion and Future Plan

This study investigated the effects of enjoyable and speculative aspects on users' playing behavior in blockchain-based games. We crawled playing behaviors of Cryptokitties and analyzed these behaviors by conducting a fixed-effect panel regression model. We found that the external aspects (i.e., the price and the total volume of Cryptocurrency) are negatively related to item-selling behavior in Blockchain-based games. In addition, the internal aspects (i.e., item purchasing, siring, gifting) are positively related to item selling. Thus, it is hard to say whether a Blockchain-based game is just a game for fun or a tool for speculation. Playing behaviors in Blockchain-based games mix enjoyable and speculative aspects. We anticipate that these findings will not only contribute to the extant literature on the speculative aspects of Blockchain-based games, but also aid game developers by suggesting how to make users continue to use the game for long-term success.

This research has some limitations. First, the characteristics of players could differ according to the time when they joined the game. For example, players who joined the game in the early stages may be more likely to be risk-takers. Unlike early adopters, players who join the game in the latter stages could be followers. Additionally, this paper does not consider the lag variables of each suggested variable. For example, the price of cryptocurrency last week could affect item selling in a Blockchain-based game this week. A later version of this paper will address these issues.

References

1. Blau, B.M.: Price dynamics and speculative trading in Bitcoin. Res. Int. Bus. Finan. **43**, 15–21 (2018)
2. Brandvold, M., et al.: Price discovery on Bitcoin exchanges. J. Int. Finan. Mark. Inst. Money **36**, 18–35 (2015)
3. CCN: Cryptocurrency market will get 'much bigger': former JPMorgan banker (2018). https://www.ccn.com/cryptocurrency-market-will-get-much-bigger-former-jpmorgan-banker/. Accessed 28 Jan 2019
4. Cheah, E.-T., Fry, J.: Speculative bubbles in Bitcoin markets? An empirical investigation into the fundamental value of Bitcoin. Econ. Lett. **130**, 32–36 (2015)
5. CNBC: Meet CryptoKitties, the $100,000 digital beanie babies epitomizing the cryptocurrency mania (2017). https://www.cnbc.com/2017/12/06/meet-cryptokitties-the-new-digital-beanie-babies-selling-for-100k.html. Accessed 28 Jan 2019
6. Coinlore. https://www.coinlore.com/. Accessed 28 Jan 2019
7. Dappradar. https://dappradar.com/category/games/. Accessed 28 Jan 2019
8. Dwyer, G.P.: The economics of Bitcoin and similar private digital currencies. J. Finan. Stab. **17**, 81–91 (2015)
9. Forbes: Five issues preventing blockchain from going mainstream: the insanely popular crypto game etheremon is one of them (2017). https://www.forbes.com/sites/outofasia/2017/12/22/five-issues-preventing-blockchain-from-going-mainstream-the-insanely-popular-crypto-game-etheremon-is-one-of-them/. Accessed 28 Jan 2019

10. Forbes: Not just Cryptokitties - the gaming industry is stepping in blockchain (2018). https://www.forbes.com/sites/yoavvilner/2018/07/10/not-just-cryptokitties-the-gaming-industry-is-stepping-in-blockchain/. Accessed 28 Jan 2019
11. Fry, J., Cheah, E.-T.: Negative bubbles and shocks in cryptocurrency markets. Int. Rev. Finan. Anal. **47**, 343–352 (2016)
12. Korzaan, M.L.: Going with the flow: predicting online purchase intentions. J. Comput. Inf. Syst. **43**(4), 25–31 (2003)
13. Liu, H.-J., Shiue, Y.-C.: Influence of facebook game players' behavior on flow and purchase intention. Soc. Behav. Pers. Int. J. **42**(1), 125–133 (2014)
14. Pilkington, M.: 11 Blockchain technology: principles and applications. In: Research Handbook on Digital Transformations, p. 225 (2016)
15. Poyser, O.: Exploring the determinants of Bitcoin's price: an application of bayesian structural time series. arXiv preprint arXiv:1706.01437 (2017)
16. González Sánchez, J.L., Padilla Zea, N., Gutiérrez, F.L.: From usability to playability: introduction to player-centred video game development process. In: Kurosu, M. (ed.) HCD 2009. LNCS, vol. 5619, pp. 65–74. Springer, Heidelberg (2009). https://doi.org/10.1007/978-3-642-02806-9_9
17. Siekpe, J.S.: An examination of the multidimensionality of flow construct in a computer-mediated environment. J. Electron. Commer. Res. **6**(1), 31–43 (2005)
18. Sweetser, P., Wyeth, P.: GameFlow: a model for evaluating player enjoyment in games. Comput. Entertainment (CIE) **3**, 3 (2005)
19. Xu, X., et al.: A taxonomy of blockchain-based systems for architecture design. In: 2017 IEEE International Conference on Software Architecture (2017)

Artificial Intelligence

You Are Not You When You Are Hungry: Machine Learning Investigation of Impact of Ratings on Ratee Decision Making

Abhishek Kathuria[(⊠)] and Prasanna P. Karhade

The University of Hong Kong, Pok Fu Lam, Hong Kong, PRC
{Kathuria, karhade}@hku.hk

Abstract. We leverage machine learning methods to investigate the role of online ratings on ratee decision making on an online food delivery platform in India. Findings reveal that in the emerging economies, ratings are not likely to have a strong bearing on certain ratee decisions on the online platform. Research on the platform economy in the emerging markets is likely to enable us to broaden our knowledge on the overall impact of online ratings.

Keywords: Platforms · Ratings · Decision tree induction · Machine learning · Emerging markets · India

1 Introduction

This past decade has witnessed rapid transformations in various industries due to the emergence of digital marketplaces and online platforms. User generated content, in the form of reviews, ratings, and comments, is a critical common characteristic across many of these marketplaces and platforms. Production of user generated content usually does not impose an explicit cost on consumers [5, 11, 24]. Instead, user generated content affects the decision-making processes of other consumers and hence many marketplaces and platforms encourage the production and curation of user generated content. Consequently, generation of content has become the primary purpose of many online platforms, such as review and rating websites like Yelp and TripAdvisor.

A substantive literature examines issues related to user generated content, its antecedents, and consequences in digital marketplaces and online platforms. Ratings, a specific type of aggregated user generated content, have been determined to especially impact strategic behavior and decision-making of other consumers. While reviews, comments and user interactions capture many nuances [24], ratings reflect all these together in a single indicator. Thus, the rating of a ratee aggregates the overall sentiment of raters and can be viewed as a codified assessment on a standardized scale [11].

Do ratings matter? This is a fundamental question that has been explored extensively. Researchers have examined the effect of three aspects of ratings: valence (e.g., [5]), variance (e.g., [3]) and volume (e.g., [5]). However, we observe a key inadequacy in the literature. Though effects of ratings on the strategic behavior and choices of raters have received extensive attention (e.g., [23]), how ratings influence the decision

© Springer Nature Switzerland AG 2019
J. J. Xu et al. (Eds.): WEB 2018, LNBIP 357, pp. 151–161, 2019.
https://doi.org/10.1007/978-3-030-22784-5_15

making of ratees has not been given equal consideration (e.g., [34]). Prior research suggest that ratings influence the decision making of ratees in the context of brand building, customer acquisition, and product development [4]. For example, research using panel data shows that sellers (ratees) with low ratings are more likely to exit eBay [2]. However, it is not necessary that ratings matter for all types of strategic decisions by ratees. Further, prior work may suffer from sample selection biases, false positives due to overfitting of data and idiosyncrasies of the marketplace being investigated. Furthermore, the influence of ratings on ratees may differ in different economic and national contexts. The combination of these factors gives rise to a gap in our under-standing regarding the efficacy of ratings on ratee decision making under specific combination of attributes. Specifically, our understanding of how ratings influence financial decisions of ratees in the context of a growing platform, in a non-western, less educated, unindustrialized, impoverished, emerging economy is limited. Formally, we aim to address the following research question:

Do ratings on an online marketplace affect the decision of the ratee to participate in financial transactions on the marketplace?

To address this research gap, we apply a machine learning classification technique on a population level dataset of restaurants, their features, ratings, and financial par-ticipation decision from a major food marketplace in India. India is one of the most diverse nations in the world, with 22 official languages, dozens of cultures and a complex gastronomic palate. It is home to over a hundred thousand restaurants in the organized sector, which serve diverse, rich, and mature cuisines. Availability of large datasets from India, combined with big data analytical techniques have contributed extensively to the emerging field of computational gastronomy (e.g., [13]). India has also been studied extensively in the management, operations management and infor-mation systems literatures (e.g., [15, 16, 37]). This paper is another step in this direction.

Our initial dataset consists of the population of nearly ninety-six thousand restaurants across 37 cities in India. After dropping restaurants without ratings, we analyzed the flow of the decision-making process of over sixty thousand restaurants by applying decision tree induction. This enabled us to model the cumulative decision experiences of the ratees and ascertain the role of ratings as the ratees go through the decision of participating in financial transactions on the marketplace [19]. Though this methodology has been used sparingly in the past (e.g., [25, 33, 35], there has been an increase in recent applications due to methodological advancements and the availability of large datasets. Decision tree induction has several advantages, including a lack of distributional assumptions and the ability to discover underlying patterns in the data and decision-making attributes [1, 17, 21]. It is especially optimal for our research question and theory development [14] due to its low rate of false positive predictions [32].

Our decision trees were grown using the C4.5 decision tree classification algorithm [29, 30]. A series of computational experiments were conducted across varied levels of pruning to uncover the role of ratings in the underlying structure of the data. We find that ratings on a platform are not part of the decision-making attributes for the rated (restaurants) when they decide whether to participate in financial transactions on the digital marketplace. In other words, we show that under specific conditions, ratings do not matter to the rated.

2 Related Literature

Related literature has demonstrated that consumers reduce their cognitive effort and resort to simplifying strategies and heuristics for decision making as a response to two issues: complexity and abundance of information; and, cognitive limitations to processing this information in limited time [11, 36]. Information that can be easily aligned or can interpreted through numeric values along a standard scale [10] is considered more accessible and less effortful to process. Thus, numerical ratings require are used by consumers to simplify (reduce) the amount of effort that they expend on making decisions regarding product selection and purchase. Formally, ratings reduce information asymmetry in digital marketplaces by soliciting and displaying information about transaction quality to market participants. Hence ratings are considered to improve market efficiency and overcome market failure [23].

There is significant related literature has examined the effect of valence [5], variance [3] and volume [5] of ratings on the decision-making process of raters with respect to product sales. A few common themes emerge from this literature. First, ratings matter, but not always: empirical results have been mixed [5, 18]. While some studies find no effect of rating variance and volume on sales, others find negative and significant effects [5]. Second, the nature of the product or service being rated and the nature of the rating system (one-sided versus two-sided ratings) matter: they influence the distribution and consequences of ratings. For example, 31% of ratings on TripAdvisor and 44% on Expedia are five-star ratings as compared to compared to 75% on Airbnb [22]. Also, some researchers have found no significant impact of ratings on box-office of movies (e.g., [5]), whereas others have found positive (e.g., [3]) and even long-term impacts (e.g., [18]). On the other hand, the positive effect of ratings on sales of electronic products has been established in several studies. Third, ratings alone are not enough: other aspects of reviews are required to explain all nuances of raters' decision-making behavior with regards to sales [24] because numeric ratings do not fully capture the polarity information in the review [6]. Thus, the effect of ratings on sales rank is mostly indirect, through sentiments, while sentiments' effect on sales rank is mostly direct [11]. Finally, on the few occasions that the effect of ratings on ratees has been examined, ratings matter. Ratees with lower ratings witness drop in sales, and more frequent subsequent lower ratings [2]. Also, low ratings increase the chance of market exits of ratees [2]. Considering these broad thematic contours of the related literature, it is plausible that ratings should affect the decision of ratees to participate in subsequent financial transactions on a digital marketplace.

3 Model Formulation

3.1 Machine Learning

Classification with machine learning, e.g., tree induction, is a data-driven methodology for discovering patterns from data [29, 30]. Induction yields easy-to-interpret, rules which shed light on tacit decision rationale to make informed inferences about decision making [1, 20]. Trees are accessible to a variety of stakeholders including top management executives and policy makers (e.g., [14, 17]) as they represent the discovered patterns in the form of a tree of if-then rules. Often, articulating business logic can be difficult for stakeholders as the underlying logic tends to be tacit.

Classification via tree induction opens the black box of the tacit business logic and represents interrelationships between various decision attributes and outcomes. Machine learning techniques for classification are effective for discovering combinations of attributes often not known ex ante, and compactly representing their cumulative influence on outcomes [17, 21]. Trees shine the light on emergent interconnections between attributes that are deemed informative (the only attributes included in the tree) [17, 21]. Thus, trees weed out features that are not informative for explaining outcomes. Moreover, tree induction makes few distributional assumptions about the data making this methodology more generalizable.

Building on the idea of partitioning, is a testing mode called n-fold validation where the data is divided into n partitions and n-1 partitions are used as the training sample and one partition (or fold) is used for validation. 10-fold validation, used in this investigation, is a popular testing mode for induction. A pitfall with analytics is that data scientists over-fit their models and explain noise in their data (as opposed to underlying relationships of interest). We take necessary precautions and not fall into the overfitting trap by using data partitioning. We assess generalizability of the knowledge discovered on training data by testing its prediction accuracy on unseen data from the validation data partition.

3.2 Inductive Model

Post data partitioning, two steps define classification via machine learning. Firstly, the C4.5 algorithm is used to grow the tree on training data [29, 30]. Secondly, tree grown in step 1 is pruned by validating it with unseen data from the validation partition. By employing high levels of pruning, we are able to discover the tacit structure of the data and demonstrate robustness of the discovered knowledge. The Weka platform, a popular, open-source platform is used for data partitioning, and for growing and pruning trees [8].

Tree induction iteratively groups together observations (i.e., restaurants) such that they are similar not only in certain information attributes, but also similar in terms of their participation in financial transactions outcomes. There are two inputs to tree induction: (1) restaurants described by all information attributes, and (2) financial transactions participation decisions. The objective of tree induction is to discover tacit combinations of information attributes associated with similar final outcomes (i.e., similar decisions regarding financial transaction participation) [29]. Trees only retain the most pertinent decision attributes for explaining decisions and organize decision attributes in a context-dependent manner; certain questions are only raised depending on answers obtained to other questions [30].

Using prediction accuracy of the decision tree as the sole criterion when choosing the best representative tree (among alternative models) can be misleading and would be akin to falling into the overfitting trap. We avoid the overreliance on the prediction accuracy by considering two other heuristics, namely communicability and consistency of the discovered knowledge. In summary, three heuristics, (i) prediction accuracy, (2) communicability, and (3) stability of the discovered knowledge guide the choice of the best representative tree.

Trees discovered by induction are not reflective of the exact rules or "scripts" used by the decision makers, but rather represent credible approximations of the decision rationale [1]. Instead of the correlations between attributes, induction relies on the amount of information an attribute conveys about the decision outcome.

3.3   Context and Data

Our research context is a large, comprehensive review and rating website based in India. This website has a pan-India presence and has been in operation for more than 2 years in all large cities in India. This website also provides a digital marketplace for food ordering. All registered restaurants in India are listed on website, irrespective of whether they participate in financial transactions in the marketplace. Thus, all restaurants receive ratings (subject to a few conditions). This effectively addresses concerns stemming from sample selection bias as we are able to observe ratees, irrespective of whether they participate in the marketplace or not. The marketplace does not levy fees from customers and thus does not cross-subsidize restaurant participation in financial transactions. Restaurants' financial transaction participation choices are therefore not influenced by dynamics of the underlying fee / payment structure. Finally, in our setting, multi-homing costs are low and a restaurant can choose to affiliate with any number of marketplaces. Research suggests that winner-take-all outcomes are unlikely in such contexts [7].

A population sample of 95,735 restaurants, serving a total of 135 different cuisines, located in the 37 cities of India form our dataset. Restaurants across India are part of the sample if they are listed on the digital marketplace. Any consumer can list a restaurant on the website; listed restaurants can garner reviews and ratings from other consumers. A strategic choice that restaurant owners must make is to choose if they wish to participate in financial transactions through the marketplace. *Mere listing does not imply participation.*

This decision is a nontrivial decision that can have different outcomes. Participating in financial transactions on the marketplace may increase demand for the restaurant's products among customers who use the marketplace. A positive outcome can be increased sales for the restaurant. However, this decision carries with it an increased risk that the restaurant may not be able to fulfill demand arising from the digital marketplace, adversely impacting its rating, and, its sales [12]. Specifically, there are three reasons for this risk. First, restaurants pay the digital marketplace a fee inversely proportional to the transaction value as per a multi-tier structure. Second, restaurants might not be able to cope with high spikes and unforeseen growth in demand. Third, adverse reputational affects can accrue owing to a mismatch in service levels at the restaurant and the stakeholders on the platform (e.g., delivery personnel).

3.4 Outcome of Interest

We investigate an individual restaurant's decision to *participate in financial transactions on the online marketplace* and thus digitize a certain proportion of their business transactions. The outcome variable, *Participation*, is coded as Yes if the restaurant participates in the financial transactions and coded as No if it does not. Next, we describe the attributes included in our theory.

3.5 Decision Attributes

We included several decision attributes. The *Cost* of a meal for two persons at the restaurant reflects the strategic positioning of the restaurant (e.g., cost leadership [26–28]). Specifically, cost for a restaurant that offers a meal for two persons for 1000 Indian Rupees (INR) and above was assigned a value of high, less than or equal to 300 INR assigned a value of low, and medium otherwise. *Number of Cuisines* was assigned a value of low if the restaurant offered a single cuisine, medium if two or three cuisines were offered. A value of high was assigned if the restaurant offered more than three cuisines. If the restaurant is a vegetarian only restaurant or not is captured by using a dummy called *Vegetarian*. Similarly, if the restaurant provides only Indian food (vs. world cuisines) is captured using a dummy called *Only Indian*. If the restaurant serves alcohol is captured using a dummy called *Alcohol*. If the restaurant provides any form of parking services is captured using a dummy called *Offers Parking*. If the restaurant provides any features that can encourage customers to dine in, such as live entertainment or music, (as opposed to ordering in) is captured using a dummy called *Go-In*. Two attributes captured a restaurant's technology readiness. First, if restaurants accept electronic payments through digital wallets was captured using a dummy called *Digi-Pay*. Second, if the restaurant provides free wi-fi internet access to its customers is captured using a dummy called *Wi-fi*.

A key institutional attribute that we captured corresponds to whether a restaurant is part of a group of restaurants with the same name. These restaurants may be part of a chain or might share a common name that reflects a well-established identity [9]. Institutional norms and processes are likely to be common across restaurants that belong to the same chain or group [31] and hence similar with regards to their propensity to participate in financial transaction on platforms. Thus, we capture this attribute by assigning *Chain* a value of high if nine or more other restaurants had the same name as the focal restaurant, medium if at least one other restaurant, and less than nine other restaurants, shared their names with the focal restaurant, and low if the restaurant's name was unique.

We also captured a key environmental attribute corresponding to the unique context of India. Restaurants located in metropolitan cities of India (Mumbai, Delhi, Chennai and Kolkata and now Pune, Hyderabad and Bangalore) are likely to be systematically different in their propensity to participate in financial transaction on online platforms compared to restaurants in the rest of India. We capture these differences using an attribute called *Metro India*.

Finally, the focal variable of our analysis, a restaurant's online *Rating* was captured. A restaurant's online rating represents its reputation or social capital in the digital world. A restaurant's offline reputation migrates to the digital marketplace as more and more customers review and rate the restaurant. Overall, since information contained in the reviews is distilled to one final online rating, we only included the overall online rating in our analysis. This website recorded a restaurant's rating on a 5-point scale. We transformed ratings from their numeric value to three categories of high, medium and low (high when greater than or equal to 4, low when less than 3, medium otherwise). Certain restaurants did not have ratings and such restaurants were excluded from our analysis.

3.6   Model Setup

To ensure that decision rationale is comprehensively discovered, a process of drawing, mutually exclusive, training and testing subsamples is repeated multiple times. An iteration of tree induction is described next. In each iteration, we draw random, mutually exclusive subsamples of restaurants from the original data; one set, known as the training set, from which the tacit decision rationale is discovered by the C4.5 induction algorithm [29], and another disjoint set of initiatives, known as the testing set, which is used to test the predictive accuracy of this discovered rationale. We used 10-fold validation where the full sample is divided into 10 partitions of which 9 partitions are used for building the tree and the last partition is used for validation. Prediction accuracy of the tree discovered from training set is assessed by predicting decisions for restaurants from unseen data from the validation set.

4   Computational Experiments

4.1   Attribute Selection and Model Identification

Multiple approximations of the tacit rationale are derived by iterating experiments where the 10-fold validation process is repeated at varying levels of pruning. These experiments are integral to induction to ensure that multiple approximations of the underlying decision process are available to the researchers. We rely on three heuristics to select the best representative, a credible approximation, of the tacit decision process: high predictive accuracy, high parsimony, and high reliability.

All twelve information attributes characterizing restaurants in conjunction with the final financial transaction participation decision, are inputs to induction. All information attributes deemed informative for explaining participation decisions are included in the trees as decision attributes and the induction algorithm excludes all the non-informative attributes from the tree. The most informative decision attribute is the top-most attribute in the tree. Importance of attributes decreases as we move away from the top of the tree to its leaves. Trees organize attributes in a context-dependent manner; certain questions are only raised depending on answers obtained to questions answered previously [30].

Table 1. Computational experiments

No.	Degree of pruning	Min instances at leaves	Number of leaves	Top two levels of decision attributes	Prediction error	Ratings
1	Low	100	25	1: Urban India 2: Digi-Pay, Cost	31.80%	Not in the tree
2	Low	200	21	1: Urban India 2: Digi-Pay, Cost	31.77%	Not in the tree
3	Low	500	24	1: Urban India 2: Digi-Pay, Cost	31.91%	Lowest in the tree
4	Medium	100	28	1: Urban India 2: Digi-Pay, Cost	27.72%	Not in the tree
5	Medium	200	21	1: Urban India 2: Digi-Pay, Cost	31.75%	Not in the tree
6	Medium	500	24	1: Urban India 2: Digi-Pay, Cost	28.12%	Lowest in the tree
7	High	100	21	1: Urban India 2: Digi-Pay, Cost	31.83%	Not in the tree
8	High	200	21	1: Urban India 2: Digi-Pay, Cost	31.83%	Not in the tree
9	High	500	19	1: Urban India 2: Digi-Pay, Cost	32.00%	Not in the tree
10	Aggressive	100	19	1: Urban India 2: Digi-Pay, Cost	31.97%	Not in the tree
11	Aggressive	200	20	1: Urban India 2: Digi-Pay, Cost	32.05%	Not in the tree
.12	Aggressive	500	20	1: Urban India 2: Digi-Pay, Cost	32.04%	Not in the tree

4.2   Experimental Setup

We generated alternative models by changing the degree of pruning and the minimum number of instances at leaves in the trees. The entire comprehensive collection of twelve attribute was used to model platform participation decisions. Across all our computational experiments, metro was consistently the top most classification attribute and the ratings attribute was absent from the decision tree.

Given our counterintuitive findings and importance of ratings in the extant literature, we explored additional combinations of the degree of pruning and the minimum number of instances (i.e., restaurants) on the leaves (see Table 1). In some scenarios, we were indeed able to induce trees which included ratings as a predictor. In all such instances, ratings were consistently the least important predictor. These findings represent strong evidence to suggest that, in this case, ratings are not critical for influencing financial participation decisions of the ratee.

4.3   Key Finding

The counter intuitive finding is that ratings is not included in the decision tree. This is a key finding from our research. Given the importance of ratings in the prior literature, this finding deserved more exploration. To accomplish this goal, we computationally modified our experimental parameters - degree of pruning and minimum number of instances at the leaves, with the purpose of further exploring the role of ratings into the decision tree. At times, though ratings did indeed appear in the decision trees, it always appeared as the lower most decision attribute. This suggests that ratings do not substantively influence the ratee to participate in financial transactions on the platform. This finding empowers us to qualify the explanatory power of ratings. Ratings are key for guiding the actions of other users on digital platforms. In some cases, ratings also guide the behavior of the ratees. In this case, user-generated ratings do not explain the financial transaction participation decisions of ratees on the food delivery platform.

5   Conclusion

In this paper, we have studied how ratings affect the strategic choices and decision making of the ratee. While the effect of ratings on the behavior of raters (e.g., consumers), has been extensively examined, to our knowledge, there have been few attempts in the literature to address this perspective [4]. Our analysis aimed to answer an important question within a specific context: do ratings on a platform affect the decision of the rated (restaurants) to participate in financial transactions on the platform? To address this research gap, we applied a machine learning classification technique on a population level dataset of restaurants, features, and ratings from a major food platform in India.

We used the C4.5 decision tree algorithm to initialize a solution on training data. We then conducted a series of computational experiments, wherein we used unseen data to repeatedly apply a 10-fold validation process at varying levels of pruning. A key advantage of this approach is while we avoid the common overfitting trap, decision trees themselves have a low rate of false positive predictions. Thus, our empirical choices enable us to qualify our key findings with high confidence. We have shown that ratings do not matter to the rated. Specifically, ratings on a digital marketplace are not part of the decision-making attributes for the rated (restaurants) when they decide whether to participate in financial transactions on the marketplace.

The findings from this study have implications for both practice and research. For practice, the implications of our findings study are two-fold. First, while ratings have been demonstrated to have a significant impact on the strategic behaviour of raters, it may not be a salient feature of the decision-making process for the ratee. Thus, for owners of digital marketplaces and online platforms, features other than ratings should be form the organizing principles for increasing participation in financial transactions and thus growing their installed base of ratees. Second, follow-up analyses can offer a nuanced view into the decision-making process of ratees regarding participation in financial transactions on non-exclusive marketplaces. Our practice implications can also extend to other contexts of non-exclusive digital marketplace participation.

For research, our work makes a key theoretical contribution. Ratings are considered a critical decision feature when studying decision making of participants of online platforms and digital marketplaces. Our study shows that ratings do not matter for specific stakeholders (the ratee), for specific decisions (participating in financial transactions), under specific contexts (growing marketplace in a non-western economy). Similar ideas should and need to be tested in other contexts and on other strategic choices made by the ratee, such as change in level of engagement, change in scope of participation, and platform abandonment. Methodologically, our use of the C4.5 decision tree algorithm, which has low rate of false positives, serves as a sample context where machine learning classification techniques can be applied [32].

Despite providing valuable insights, our results must be interpreted within the boundaries of the study. As noted, an interesting extension of this research would be to incorporate different types of strategic choices as the consequence of ratings. Another limitation is the generalizability of our results to other contexts (online platforms and digital marketplaces) may be limited. This is an interesting scope for future research studies.

References

1. Boonstra, A.: Structure and analysis of IS decision-making processes. Eur. J. Inf. Syst. **12**(3), 195–209 (2003)
2. Cabral, L., Hortacsu, A.: The dynamics of seller reputation: evidence from eBay. J. Ind. Econ. **58**(1), 54–78 (2010)
3. Chintagunta, P.K., Gopinath, S., Venkataraman, S.: The effects of online user reviews on movie box office performance: accounting for sequential rollout and aggregation across local markets. Mark. Sci. **29**(5), 944–957 (2010)
4. Dellarocas, C.: The digitization of word of mouth: promise and challenges of online feedback mechanisms. Manag. Sci. **49**(10), 1407–1424 (2003)
5. Duan, W., Gu, B., Whinston, A.B.: Do online reviews matter?—an empirical investigation of panel data. Decis. Support Syst. **45**(4), 1007–1016 (2008)
6. Ghose, A., Ipeirotis, P.G.: Estimating the helpfulness and economic impact of product reviews: mining text and reviewer characteristics. IEEE Trans. Knowl. Data Eng. **23**(10), 1498–1512 (2011)
7. Hagiu, A.: Two-sided platforms: product variety and pricing structures. J. Econ. Manag. Strategy **18**(4), 1011–1043 (2009)
8. Hall, M., Frank, E., Holmes, G., Pfahringer, B., Reutemann, P., Witten, I.H.: The WEKA data mining software: an update. ACM SIGKDD Explor. Newslett. **11**(1), 10–18 (2009)
9. Hannan, M.T., Freeman, J.: The population ecology of organizations. Am. J. Sociol. **82**(5), 929–964 (1977)
10. Hsee, C.K.: The evaluability hypothesis: an explanation for preference reversals between joint and separate evaluations of alternatives. Organ. Behav. Hum. Decis. Process. **67**(3), 247–257 (1996)
11. Hu, N., Koh, N.S., Reddy, S.K.: Ratings lead you to the product, reviews help you clinch it? The mediating role of online review sentiments on product sales. Decis. Support Syst. **57**, 42–53 (2014)
12. India, N.R.A.I.o.: Food tech startups leave bad taste in restaurants' mouths (2016)

13. Jain, A., Rakhi, N.K., Bagler, G.: Analysis of food pairing in regional cuisines of India. PLoS One **10**(10), e0139539 (2015)
14. Karhade, P., Shaw, M.J., Subramanyam, R.: Patterns in information systems portfolio prioritization: evidence from decision tree induction. MIS Q. **39**(2), 413–433 (2015)
15. Kathuria, A., Mann, A., Khuntia, J., Saldanha, T., Kauffman, R.J.: A strategic value appropriation path for cloud computing. J. Manag. Inf. Syst. **35**(3), 740–775 (2018)
16. Kathuria, R., Porth, S.J., Kathuria, N., Kohli, T.: Competitive priorities and strategic consensus in emerging economies: evidence from India. Int. J. Oper. Prod. Manag. **30**(8), 879–896 (2010)
17. Langley, A., Mintzberg, H., Pitcher, P., Posada, E., Saint-Macary, J.: Opening up decision making: the view from the black stool. Organ. Sci. **6**(3), 260–279 (1995)
18. Lee, Y.-J., Hosanagar, K., Tan, Y.: Do I follow my friends or the crowd? Information cascades in online movie ratings. Manag. Sci. **61**(9), 2241–2258 (2015)
19. March, J.G.: Primer on Decision Making: How Decisions Happen. Simon and Schuster, New York (1994)
20. March, J.G., Shapira, Z.: Managerial perspectives on risk and risk taking. Manag. Sci. **33**(11), 1404–1418 (1987)
21. Markus, M.L., Majchrzak, A., Gasser, L.: A design theory for systems that support emergent knowledge processes. MIS Q. **26**, 179–212 (2002)
22. Mayzlin, D., Dover, Y., Chevalier, J.: Promotional reviews: an empirical investigation of online review manipulation. Am. Econ. Rev. **104**(8), 2421–2455 (2014)
23. Pallais, A.: Inefficient hiring in entry-level labor markets. Am. Econ. Rev. **104**(11), 3565–3599 (2014)
24. Pavlou, P.A., Dimoka, A.: The nature and role of feedback text comments in online marketplaces: implications for trust building, price premiums, and seller differentiation. Inf. Syst. Res. **17**(4), 392–414 (2006)
25. Pomerol, J.-C., Brézillon, P., Pasquier, L.: Operational knowledge representation for practical decision-making. J. Manag. Inf. Syst. **18**(4), 101–115 (2002)
26. Porter, M.E.: Competitive Strategy: Techniques for Analyzing Industries and Competition, vol. 300, p. 28. Simon and Schuster, New York (1980)
27. Porter, M.E.: Competitive Advantage: Creating and Sustaining Superior Performance. Free Press, New York (1985)
28. Porter, M.E., Millar, V.E.: How Information Gives You Competitive Advantage. Harvard Business Review, Brighton (1985)
29. Quinlan, J.R.: Induction of decision trees. Mach. Learn. **1**(1), 81–106 (1986)
30. Quinlan, J.R.: Decision trees and decision-making. IEEE Trans. Syst. Man Cybern. **20**(2), 339–346 (1990)
31. Scott, W.R.: The adolescence of institutional theory. Adm. Sci. Q. **32**, 493–511 (1987)
32. Spangler, W.E., May, J.H., Vargas, L.G.: Choosing data-mining methods for multiple classification: representational and performance measurement implications for decision support. J. Manag. Inf. Syst. **16**(1), 37–62 (1999)
33. Tessmer, A.C., Shaw, M.J., Gentry, J.A.: Inductive learning for international financial analysis: a layered approach. J. Manag. Inf. Syst. **9**(4), 17–36 (1993)
34. Tiwana, A.: Platform desertion by app developers. J. Manag. Inf. Syst. **32**(4), 40–77 (2015)
35. Tsang, E., Yung, P., Li, J.: EDDIE-Automation, a decision support tool for financial forecasting. Decis. Support Syst. **37**(4), 559–565 (2004)
36. Tversky, A., Kahneman, D.: Judgment under uncertainty: heuristics and biases. Science **185**(4157), 1124–1131 (1974)
37. Venkatesh, V., Shaw, J.D., Sykes, T.A., Wamba, S.F., Macharia, M.: Networks, technology, and entrepreneurship: a field quasi-experiment among women in rural India. Acad. Manag. J. **60**(5), 1709–1740 (2017)

Ensemble Classification Method
for Imbalanced Data Using Deep Learning

Yoon Sang Lee[(⊠)]

Columbus State, Columbus, GA 31909, USA
lee_yoon@columbusstate.edu

Abstract. Nowadays various types of devices provide abundant data, and many businesses want to pinpoint what they are interested in from the data such as target marketing, fraud transaction detection. However, current classification algorithms in data mining show a poor performance when classifying imbalanced data.

To enhance the classification performance of minority class in imbalanced datasets, we present an ensemble learning method using the combination of an UnderBagging, a majority vote, and a meta classifier giving higher decision priority to the classifier that predicts a class better, basing on Deep Neural Network (DNN) as a classifier. We tested our method with two imbalanced datasets from UCI Data Repository and compared the performance of our method with four other techniques. The result showed that our method provided an improved performance on classifying minority class instances compared to the other four techniques.

Keywords: Class imbalance · Bagging · Under-sampling · Ensemble · Deep Neural Network

1 Introduction

In recent years, various types of IT devices provide abundant data, and many enterprises want to pinpoint the target in which they are interested to reduce the effort and cost. For example, businesses want to find out fraud transactions from normal ones in credit card fraud detection or want to identify a small portion of customers from a large customer population when sending out mails to target customers. However, the performance of most of predictive machine learning or data mining algorithms deteriorates when dealing with skew datasets since they assume balanced datasets [1]. In class imbalance literature, various studies have suggested methods to improve the classification performance of minority class such as over/under-sampling, classifier ensemble, or new cost-functions. However, there still exists a gap to improve the performance of imbalanced data classification. Hence, we propose a DNN-based model in this working paper. We trained an ensemble of DNNs by applying the UnderBagging and a majority vote to build a classifier that identifies the minority class examples better. We tested our method with two imbalanced datasets and compared the performance with four other techniques using C4.5 classifier. The result showed that our method provided a better performance than other techniques in the prediction of minority class.

© Springer Nature Switzerland AG 2019
J. J. Xu et al. (Eds.): WEB 2018, LNBIP 357, pp. 162–170, 2019.
https://doi.org/10.1007/978-3-030-22784-5_16

2   Literature Review

For the class imbalance problem, there have been two categories of approaches: (1) Sampling-based approach and (2) Algorithm-based approach. Sampling-based approach focuses on the data itself to enhance the classification performance. Over-sampling method [2–4] creates a balanced dataset by duplicating the minority examples in imbalanced dataset. This method is effective, but might cause a overfitting because of duplicated samples [5]. Under-sampling approach [3, 6] creates multiple balanced subsets by resampling the majority examples from the imbalanced dataset. It is empirically known that it enhances the performance of classification, but has a possibility of missing the important information in unselected majority examples [1]. SMOTE [7, 8] is a type of over-sampling approach, but it synthesizes new minority instances by using k-nearest neighbor algorithm and interpolating them. Algorithm-based approach tries to mitigate the class imbalance problem by enhancing the performance of classification algorithm. In one-class learning method [9–11], a classifier is trained by the dataset having one class examples. Cost-sensitive approach [12–14] is another algorithm-based approach, and the strategy of this approach is to minimize the cost of misclassification by imposing a higher cost to false negative than false positive.

In addition to the studies using a single classifier, ensemble of classifiers is applied to the class imbalance problem. SMOTEDBoost [15] and MSOTEDBoost [16] are proposed. Basically, they used AdaBoost, but the instances synthesized by SMOTE or MSMOTE are added to the training instances. Another group of studies used a bagging ensemble for class imbalance problem. OverBagging uses a bagging algorithm and it creates bags by sampling and replicating minority instances. SMOTEBagging [17] creates bags by combining resampled minority instances and minority instances synthesized by SMOTE. Random balance [18] also synthesizes new samples using SMOTE, but the proportion of the classes in each training set for each classifier is decided randomly. UnderBagging [19] under-samples from a majority class set or from a minority class set with replacement to create balanced bags. [1] created multiple disjointed training sets and trained a classifier for each training set.

The studies on class imbalance were conducted in many domains. Especially, the studies of class imbalance techniques in business area appeared in information system and management journals to proposed solutions for imbalance data sets in finance and business management [3, 20] used oversampling for customer churn prediction, and [21] suggested a support vector machine based CRM model for imbalanced data. [22] proposes a nearest neighbor classification model to deal with the class imbalance and a convex optimization technique, and [23] used a support vector machines with soft margins as the base classifier to counteract the imbalance data set. [24] used a Ada-Boost and neural networks to build a bankruptcy forecasting model. [25] used a undersampling technique in opinion mining within media analysis since there exist an imbalance in the distribution of positive and negative samples.

While class imbalance studies did not consider a specific type of classifier or generally used C4.5 as a classifier, Deep Learning [26] has gained the popularity in academia and industry due to its enhanced classification performance. However, the applications of DNN to structured data were rare due to the limitation of input data format. One-hot encoding has been considered as an alternative, but it was not effective in many cases. However, [27] and [28] proposed an entity embedding as the alternative for the problem. Although the entity embedding is suggested for the classification of structured data, to the best of our knowledge, Deep Learning is not applied to the structured imbalance dataset.

3 Method

In this study, we proposed a new method using the combination of an UnderBagging, a majority vote, and a meta classifier giving higher decision priority to the classifier that predicts a class better. Figure 1 depicts the method proposed in the study. For the model training, first we separate the entire training set into majority instances and minority instances. Then, we randomly resample the same number of majority instances with the number of minority instances from the majority instances without replacement and combined them with minority instances to form a new balanced bag. In this manner, we may create multiple balanced bags by repeatedly resampling majority instances and combining the selected majority instances with minority instances. Since the high variety of training sets increase the performance of ensemble classification [29], we decided the number of bags as shown in Eq. (1)

$$\text{The number of bags} = \text{Ceil}\,(\text{majority instances/minority instances}) + 2 \qquad (1)$$

The formula is devised to include all the information in all instances by providing enough number of subsets that could stochastically include any majority instances at least once when they are resampled to introduce enough variety. In the preliminary

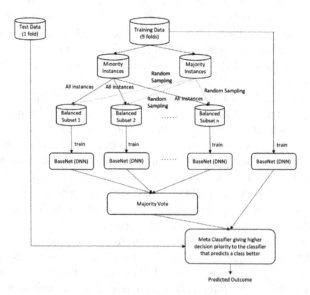

Fig. 1. Ensemble learning using UnderBagging, majority vote, and a meta classifier

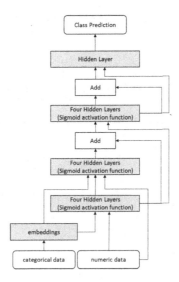

Fig. 2. Structure of BaseNet

experiments, we found that the performance of method has improved as we increase the number of bags until reaching the number of bags in Eq. (1), but there was no further performance improvement with the larger number of bags than that in Eq. (1).

Once multiple balanced bags are created, each bag is trained by the same type of classifier. The classifier of this model is created by using DNN (entitled BaseNet). Since a DNN model showed a better performance on the detection of minority instances in our preliminary experiments when it is applied to the ensemble model proposed by [1], we selected DNN as the classifier of our ensemble model. The DNN for our model is designed to be able to provide a higher prediction accuracy for majority instances when an imbalanced set is provided as an input and to provide higher accuracy for minority when a balanced set is used as an input.

The BaseNet is composed of 13 linear layers and an entity embedding is used for categorical variables. Also, concatenation and addition layers are used in the middle of linear layers to avoid vanishing gradient problem, which is an approach used in [30] for image processing, but it is used for structured data in our study. The structure of BaseNet is depicted in Fig. 2. The results from multiple BaseNets are combined by majority vote to form a classifier providing higher prediction accuracy for minority instances. Apart from balanced bags, another BaseNet is trained by the entire training set. Note that the entire training set is an imbalanced set. The result of this BaseNet provides a higher prediction accuracy for majority instances. Once two types of classifiers, one for high majority prediction accuracy and another for higher minority prediction accuracy, are trained, the results from two classifiers are combined into a meta classifier to make a final decision for the entire classification. The meta classifier provides higher decision priority to two types of classifiers in a sequential way for the classification of the instance. If two classifiers cannot identify the classifier for which they have strength, the decisions of two classifiers are combined by applying the weight of each classifier. Details of meta classifier's algorithm are in Table 1.

Table 1. Algorithm used for a meta classifier

```
W_maj = weight of classifier for major;
W_min = weight of classifier for minor;

If the result of classifier for minor is a posi-
tive class{
     Final decision is positive class;
}
else{
     If the result of classifier for major is a
negative class
          Final decision is negative class;
     else{
          Final decision is Maximum(the result of
classifier for minor * W_min ,
                              the result of
classifier for major * W_maj);
     }
}
```

For the evaluation of the model, 10-fold cross-validation is conducted by using two datasets from UCI Machine Learning Repository (https://archive.ics.uci.edu/ml/index.php). Table 2 provides the details of the datasets. The model of this study is implemented by using Pytorch Deep Learning library. In addition to our model, four classification methods suggested for class imbalance problem are implemented and the results are compared. As the based line of comparison, we conducted 10-fold cross-validation with the original dataset and C4.5 algorithm. As the second classification method, we applied SMOTE to the dataset and conducted 10-fold cross-validation with C4.5 algorithm. As the third, we applied over-sampling to the dataset and analyzed the over-sampled dataset with C4.5 algorithm. As the last one, we implemented the method suggested in [1] and conducted in 10-fold cross-validation since this approach used the same bagging and majority voting strategy to the one made in this study. For these methods, we used J48 algorithm in WEKA machine learning package.

Table 2. Datasets used for experiments

Data set	The number of attributes	The number of instances	Majority percentage
Adult	14	32,560	75.9%
Bank marketing	20	41,188	88.7%

4 Results and Discussion

Table 3 shows the results from five classification methods. In the results, four class imbalance methods showed the drops of accuracy from the baseline result. Among them, our methods showed the largest accuracy drops. However, this is usually considered a trade-off in dealing with class imbalance problem, and this amount can be considered not significant performance drops [1].

Table 3. Total accuracy of the methods

Techniques Dataset	Original dataset	SMOTE	Over-sampling	The method in Sikora and Rania [1]	Our method
Adult	86%	85%	82%	79%	77%
Bank marketing	91%	90%	89%	85%	80%

Although the total accuracy is an important measure to evaluate the performance of an algorithm for balanced datasets, using the prediction accuracy might be misleading since, for example, even predicting only minority class can yield a high prediction accuracy. Hence, we compare the prediction accuracy of each class and F-Score because the goal of class imbalance study is to improve the prediction accuracy of minority instances. The results are shown in Table 4.

Table 4. Prediction accuracy of each class

Techniques Dataset Dataset		Original dataset	SMOTE	Over-sampling	The method in Sikora and Rania [1]	Our method
Adult	Majority (Specificity)	94%	91%	84%	76%	72%
	Minority (Sensitivity)	63%	67%	76%	89%	92%
	Difference	31%	24%	8%	13%	20%
	F-Score	0.604	0.685	0.669	0.668	0.653
Bank marketing	Majority (Specificity)	96%	93%	91%	84%	77%
	Minority (Sensitivity)	54%	66%	73%	94%	97%
	Difference	42%	27%	18%	10%	20%
	F-Score	0.533	0.586	0.603	0.589	0.517

In Table 4, we also could observe the trade-off between sensitivity and specificity. In majority class prediction, the methods showed the best performance in the order of: (1) Original dataset, (2) SMOTE, (3) Over-Sampling, (4) The method in [1], and (5) our method in both datasets. Our method also showed the lowest F-Score among four methods even if the differences with other methods are within 0.03 and 0.08 in two datasets. However, in minority class prediction, our method provided the best performance. Also, in the gap analysis between majority class prediction and minority class prediction, our method placed in the middle among five methods. The result implies that our method could detect the cases of interest better than other methods while maintaining a similar algorithm performance to those of others, but, at the same time, will create more false alarms. When considering the situation related to imbalance problems such as credit card fraud detection or target marketing, the cost of losing the targeted instances is much larger than that of creating false positive alarms. Therefore, when considering the highest minority prediction accuracy and the cost of losing the instances of interest, we may conclude that our method provided an improved performance for minority class prediction.

5   Conclusion and Future Work

In the paper, we proposed a new method using the combination of UnderBagging, a majority vote, and a meta classifier giving higher decision priority to the classifier that predicts a class better. We used a DNN to create a base classifier and trained them separately with multiple balanced datasets and an imbalanced dataset. We conducted experiments with four other methods, and our method provided the best performance for minority class prediction while showing the worst performance for majority class prediction. However, this is considered the trade-off between sensitivity and specificity. When considering the purpose of imbalance dataset classification and the cost of missing the target instances that we are interested in, one may conclude that our method improves the prediction performance for minority instances. The contribution of the study is that our method improves the performance of imbalance data analysis and showed that DNN can be used to solve class imbalance problem. In practice, we expect that this method can be applied to the business areas requesting the detection of minority group from larger majority such as fraud transaction detection, malicious networking traffic detection, rare disease detection, detection of fraudulent telephone calls, information retrieval and filtering tasks, and so on [1]. Since this study is in research-in-progress, we will conduct the experiment to test if the method in the study could improve the performance for the small datasets and highly skewed the datasets to generalize the result.

References

1. Sikora, R., Rania, S.: Controlled under-sampling with majority voting ensemble learning for class imbalance problem. In: Proceedings of the IEEE Computing Conference, London, UK (2018)

2. Batista, G.E.A.P.A., Prati, R.C., Monard, M.C.: A study of the behavior of several methods for balancing machine learning training data. SIGKDD Explor. Newsl. **6**, 20–29 (2004)
3. Haixiang, G., Yijing, L., Shang, J., Mingyun, G., Yuanyue, H., Bing, G.: Learning from class-imbalanced data: review of methods and applications. Expert Syst. Appl. **73**, 220–239 (2017)
4. Levi, G., Hassncer, T.: Age and gender classification using convolutional neural networks. In: 2015 IEEE Conference on Computer Vision and Pattern Recognition Workshops (CVPRW), pp. 34–42 (2015)
5. Buda, M., Maki, A., Mazurowski, M.A.: A systematic study of the class imbalance problem in convolutional neural networks. Neural Netw. **106**, 249–259 (2018)
6. Drummond, C., Holte, R.C.: C4. 5, class imbalance, and cost sensitivity: why under-sampling beats over-sampling. In: Workshop on Learning from Imbalanced Datasets II, pp. 1–8. Citeseer (2003)
7. Chawla, N.V., Bowyer, K.W., Hall, L.O., Kegelmeyer, W.P.: SMOTE: synthetic minority over-sampling technique. J. Artif. Intell. Res. **16**, 321–357 (2002)
8. Han, H., Wang, W.-Y., Mao, B.-H.: Borderline-SMOTE: a new over-sampling method in imbalanced data sets learning. In: Huang, D.-S., Zhang, X.-P., Huang, G.-B. (eds.) ICIC 2005. LNCS, vol. 3644, pp. 878–887. Springer, Heidelberg (2005). https://doi.org/10.1007/11538059_91
9. Japkowicz, N., Stephen, S.: The class imbalance problem: a systematic study. Intell. Data Anal. **6**, 429–449 (2002)
10. Kowalczyk, A., Raskutti, B.: One class SVM for yeast regulation prediction. SIGKDD Explor. Newsl. **4**, 99–100 (2002)
11. Raskutti, B., Kowalczyk, A.: Extreme re-balancing for SVMs: a case study. SIGKDD Explor. Newsl. **6**, 60–69 (2004)
12. Domingos, P.: MetaCost: a general method for making classifiers cost-sensitive. In: Proceedings of the Fifth ACM SIGKDD International Conference on Knowledge Discovery and Data Mining, pp. 155–164. ACM, New York (1999)
13. Elkan, C.: The foundations of cost-sensitive learning. In: Proceedings of the 17th International Joint Conference on Artificial Intelligence – Volume 2, pp. 973–978. Morgan Kaufmann Publishers Inc., San Francisco (2001)
14. Pazzani, M.J., Merz, C.J., Murphy, P.M., Ali, K.M., Hume, T., Brunk, C.: Reducing misclassification costs. In: Proceedings of the Eleventh International Conference on International Conference on Machine Learning, pp. 217–225. Morgan Kaufmann Publishers Inc., San Francisco (1994)
15. Chawla, N.V., Lazarevic, A., Hall, L.O., Bowyer, K.W.: SMOTEBoost: improving prediction of the minority class in boosting. In: Lavrač, N., Gamberger, D., Todorovski, L., Blockeel, H. (eds.) PKDD 2003. LNCS (LNAI), vol. 2838, pp. 107–119. Springer, Heidelberg (2003). https://doi.org/10.1007/978-3-540-39804-2_12
16. Hu, S., Liang, Y., Ma, L., He, Y.: MSMOTE: improving classification performance when training data is imbalanced. In: 2009 Second International Workshop on Computer Science and Engineering. WCSE 2009, pp. 13–17. IEEE (2009)
17. Wang, S., Yao, X.: Diversity analysis on imbalanced data sets by using ensemble models. In: 2009 IEEE Symposium on Computational Intelligence and Data Mining. CIDM 2009, pp. 324–331. IEEE (2009)
18. Díez-Pastor, J.F., Rodríguez, J.J., García-Osorio, C., Kuncheva, L.I.: Random balance: ensembles of variable priors classifiers for imbalanced data. Knowl.-Based Syst. **85**, 96–111 (2015)
19. Barandela, R., Valdovinos, R.M., Sánchez, J.S.: New applications of ensembles of classifiers. Pattern Anal. Appl. **6**, 245–256 (2003)

20. Verbeke, W., Dejaeger, K., Martens, D., Hur, J., Baesens, B.: New insights into churn prediction in the telecommunication sector: a profit driven data mining approach. Eur. J. Oper. Res. **218**, 211–229 (2012)
21. Lessmann, S., Voß, S.: A reference model for customer-centric data mining with support vector machines. Eur. J. Oper. Res. **199**, 520–530 (2009)
22. Ando, S.: Classifying imbalanced data in distance-based feature space. Knowl. Inf. Syst. **46**, 707–730 (2016)
23. Wang, B.X., Japkowicz, N.: Boosting support vector machines for imbalanced data sets. Knowl. Inf. Syst. **25**, 1–20 (2010)
24. Lane, P.C., Clarke, D., Hender, P.: On developing robust models for favourability analysis: model choice, feature sets and imbalanced data. Decis. Support Syst. **53**, 712–718 (2012)
25. Alfaro, E., García, N., Gámez, M., Elizondo, D.: Bankruptcy forecasting: an empirical comparison of AdaBoost and neural networks. Decis. Support Syst. **45**, 110–122 (2008)
26. LeCun, Y., Bengio, Y., Hinton, G.: Deep learning. Nature **521**, 436 (2015)
27. Guo, C., Berkhahn, F.: Entity embeddings of categorical variables. arXiv preprint arXiv: 1604.06737 (2016)
28. De Brébisson, A., Simon, É., Auvolat, A., Vincent, P., Bengio, Y.: Artificial neural networks applied to taxi destination prediction. arXiv preprint arXiv:1508.00021 (2015)
29. Galar, M., Fernandez, A., Barrenechea, E., Bustince, H., Herrera, F.: A review on ensembles for the class imbalance problem: bagging-, boosting-, and hybrid-based approaches. IEEE Trans. Syst. Man Cybern. Part C (Appl. Rev.) **42**, 463–484 (2012)
30. He, K., Zhang, X., Ren, S., Sun, J.: Deep residual learning for image recognition. In: Proceedings of the IEEE Conference on Computer Vision and Pattern Recognition, pp. 770–778 (2016)

Color Trend Forecasting with Emojis

Wenwen Li$^{(\boxtimes)}$ and Michael Chau

The University of Hong Kong, Pokfulam Road, Pok Fu Lam, Hong Kong
liwwen@connect.hku.hk, mchau@business.hku.hk

Abstract. Color trends are fickle components of clothing styles. It's a tough task to predict trendy colors for the fashion industry. Meanwhile, excess inventory of certain colors and stock out of popular colors both lead to extra costs. Intense competition and short product life cycles require fashion apparel retailers to be flexible and responsive to the change of market trends. As a consequence of limited historical data, many studies focus on employing advanced and hybrid models to improve forecasting accuracy. These studies ignore abundant user interaction data on social media, which is an important source to understand consumer need, as well as advanced methods to deal will multivariate data in the forecasting model. Thus, this study aims to fill this research gap by applying Bayesian Neural Networks model and incorporating user interaction data, especially emojis, into the model. The evaluation results show that Bayesian Neural Networks outperform baseline model (Neural Networks and Support Vector Regression) and the model with emoji performs better than the one without emoji. The paper demonstrates the predictive value of emoji and provides an advanced method to process multivariate data.

Keywords: Emojis · Bayesian Neural Networks · Color trend

1 Introduction

As retail competition in the fashion industry around the world continues to intensify, fashion apparel firms have to improve their management abilities that support marketing strategies and decisions [1]. Retail inventory management is a fundamental part of management operations, which heavily relies on deep understanding of market trends. Fashion retailing industry mainly includes two types of market trends, which are color trend and style trend. Color trend forecasting is an important dimension of understanding customer needs [2] and avoiding unnecessary inventory cost.

Color trend forecasting relates closely to sales forecasting but heavily depends on market information and customer preference. Even if the product is well designed, its unpopular color will prevent potential customers from making a purchase. Despite the attention that retailers have devoted to color trend forecasting, the field of information system has not yet studied the opportunities that an integrated system incorporating abundant external information offers to improve efficiency and accuracy of color trend forecasting. In this study, we develop an integrated system that applies internal transaction information from the retailer and external market information from social media to predict color trend for retailers.

© Springer Nature Switzerland AG 2019
J. J. Xu et al. (Eds.): WEB 2018, LNBIP 357, pp. 171–181, 2019.
https://doi.org/10.1007/978-3-030-22784-5_17

There are two challenges that may baffle many retailers and researchers during color trend prediction. The first is to obtain the market and consumer data and decide valuable information. Color trend is a kind of collective selections by consumers. Besides inner operational data, market-related and consumer-related data are important for improving the accuracy of prediction. To predict color trends is a tough work because of all the uncertainties associated with industries and customer tastes. Color trend forecasting depends on drawing on market information from multitudes of industries like brands, designers, magazines and celebrities. Social media provide firms and experts a platform to present their ideas on fashion products and color trends. On the other hand, customer preference has great impact on color trends, and customers can interact with individuals in fashion industry on social media. Therefore, we collaborate with experts and deliberately design the data collection process to extract highly-valued information from social media. User interactions on Facebook can be separated into two categories: one is general interactions, such as "Likes", "Comment", and "Share"; another one is reactions (including "Love", "Haha", "Wow", "Sad", and "Angry") that offers users more choices than "Likes" to directly show their feeling for the post. Reactions are line-up of emojis, and we directly call them emoji in this paper. We categorize "Likes" as a general interaction because Facebook Likes are very popular [3]. It contains richer meaning than simple emojis. General interactions have played an important role in online marketing campaign. However, emojis as a relative new function on social media do not draw much attention.

The second challenge is efficiently incorporating limited operational data and abundant external data into an integrated predictive system. Offline sales often generate daily or weekly transactional data. While various information is available on the social media platforms, monthly or weekly operational data decides a small sample size and results a relatively large dimensional dataset. Dataset with too many features can easily make machine learning methods overfitting and influence the accuracy of forecasting. However, reducing features is not an ideal solution because it will lose some information. Different from medical and astronomical data, the number of features in business data is not tremendous, but most features are valuable for decision making. Features extracted from social media data are also convey different information. Hence, we choose a variation of standard Neural Networks, which is Bayesian Neural Network (BNNs), to incorporate various features. Because of its specific way to represent parameters, BNNs can efficiently avoid overfitting problem.

The aim of this paper is twofold: on the one hand, we investigate the operational and predictive value of user interaction data, especially emojis, in a context of the fashion industry. On the other hand, we compare the performance of three main machine learning methods on multivariate data and demonstrate the advantage of Bayesian Neural Networks of avoiding overfitting problem.

2   Research Background

Researchers have developed the measurement and system for fashion color in product designing [4, 5]. Before product designing, companies need to understand color trends in the market and try to predict the future trend. Some research on color trend forecasting has been conducted in textile field [2, 6]. There are two streams of literature that are relevant to our work: the literature related to emojis and the literature on forecasting methods in fashion industry.

2.1 Emojis

Forecasting in fashion industry requires deep understanding of the market and consumers. Operational data provided by firms have been widely applied in previous studies, whereas it covers very limited information about the market and consumers. Thus, other data sources are needed. Social media data has been widely adopted as one of the most important and ubiquitous sources to reach consumers and the market. User interactions on social media has become valuable information for marketing and operation management. Besides general interaction methods (e.g. comment, share, and "Likes"), emoji is an interesting way for users to express their feeling.

Emojis are "picture characters" or pictographs [7] that can be inserted into electronic communication platforms, such as text messages, email messages and social media posts. Due to emoji's popularity and broad usage, the Oxford Dictionary named 2015 the year of emoji [8]. The main objective of emojis is to convey conversational context, like happiness, sadness, frustration, sarcasm, etc. Emojis are essentially fulfilling the function of nonverbal cues in spoken communication [9]. Besides general sentiment expression, emojis have become a popular way for consumers to express their opinions and purchase intentions. Therefore, it is feasible to detect consumers' opinion through emojis that are used in reviews and posts. In a food context, emojis are an easy and intuitive way to express emotions on meals, and consumers spontaneously express food-related emotions in tweets [10].

In this paper, we focus on Facebook reactions that are expressed as emojis including five different animated emotions: Love, Haha, Wow, Sad, and Angry. With reactions, people can react to posts in a more accurate way. Before 2016, there was only the "Like" button on Facebook. However, a thumbs-up may not be an appropriate choice for every post. The appearance of reactions allows people to express their nuanced sentiments to posts. The reactions feature has become popular among Facebook users. On February 24th, 2017, Facebook revealed that people shared 300 billion reactions at the one year mark of the feature launch. Reactions can drive user engagement and reflect people's opinions to posts. Companies are able to learn the preference of customers by their reactions record. In addition, Facebook weights reactions more than "Likes" to determine the sequence of posts on user's News Feed. Facebook reactions provide opportunities for more precise sentiment analysis. A variety of different reactions represent people's emotional responses and are much easier to analyze than textual contents. In this paper, we use the word "emoji" instead of "reaction".

2.2 Forecasting Methods in Fashion Industry

The predicting methods applied in fashion studies differ widely from traditional statistical methods to machine learning techniques. One of the earliest and useful methods is exponential model, which uses the exponentially weighted moving average to forecast the expected value of a stochastic variable [11]. The advantage of this method is requiring little information storage and slight time to compute and making relative accurate forecasts. Subsequent studies examined and extended this method, especially in the field of sales prediction [12]. Other traditional statistical methods include linear regression and auto regression integrated moving average (ARIMA), which are good at analyzing time series data. However, real data in the company may not be perfect time series data and contains much noises, and these traditional statistical methods are unlikely to perform well.

Therefore, some more complicated and powerful methods, such as machine learning, have been adopted [13, 14]. When there is an absence of theory to guide model identification, machine learning methods, such as neural networks (NN) and support vector machines (SVM), can provide accurate prediction. Comparison between NN and statistical based models has been conducted and the result shows that NN outperforms in fashion retail sales forecasting [15]. Besides NN, SVM and random forest (RF) are also important methods in fashion prediction. SVM, a semiparametric technique, performs well on predictive tasks where the relationship between predictors and target is complex [16]. SVM has one variant call support vector regression (SVR), which is an advanced method for predicting. The development of this rich class of nonlinear models inspired researchers and organizations to dig large amounts of internal and external data and obtain much more accurate prediction results [17]. Bayesian Neural Network (BNN) is a combination of probabilistic model and a neural network. In 1987, researchers have come up with an idea of Bayesian integration over network parameters [18]. BNN can be regarded as an extension of standard networks with posterior inference. Because of incorporating Bayesian inference, BNN can avoid overfitting problem.

With the advent of big data, firms always encounter large amount of external data. One the other side, the core operational data may be very limited. Machine learning methods are fit to do large dimensional or multivariate data modelling. However, when facing large dimensional data, overfitting is hard to avoid and thus causes poor performance. One kind of popular methods to process large dimensional data is variable reduction. For instance, correlation analysis, principal component analysis, and chi-square are widely applied in Information System studies. Nevertheless, dimension reduction suffers from information loss and low accuracy.

In this paper, we use BNN to overcome this problem. Standard neural network with backpropagation has some disadvantages, such as many hyperparameters that require specific tuning and a tendency to overfitting [19]. Using Bayesian inference to learning neural networks can avoid these disadvantages. BNN has been proved to be a good choice to prevent overfitting when data is scarce [20, 21].

3   BNN and Forecasting Framework

3.1   Forecasting Framework

The forecasting framework includes both operational information and social media information. Social media information is extracted from Facebook posts and consists of general interactions and emojis. Sales data provides operational information. As shown

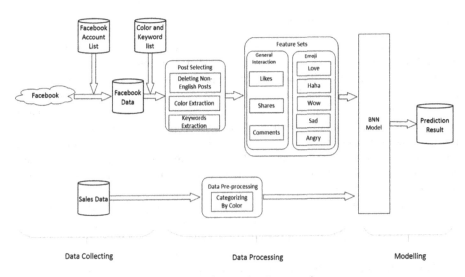

Fig. 1. The forecasting framework.

in Fig. 1, our framework consists of three phrases: data collection, data processing, and modelling.

In the first phrase, we built a Facebook account list according to the criteria we will introduce in Data Section and collect posts of all accounts on the list using Facebook application programming interface (API). Since we focus on fashion industry, four types of accounts are considered: brands, magazines, designers, and fashion influencers. Then we cleaned the data by removing hash tags, non-English characters and URLs in the text. Although accounts are related to fashion, we cannot guarantee that all posts published by these accounts are about fashion and color. To reduce noise, we built a color dictionary and a keyword dictionary with experts' help and selected posts that contain words in both of these two dictionaries. The color dictionary included all color words that can be used to describe clothing color. We categorized these color words into general nine color groups: White, Black & Gray, Blue, Green, Yellow, Red, Brown, Purple, and Orange. The keyword dictionary consisted of words that are related to clothing and fashion. There are 954 words in color dictionary and 870 words in keyword dictionary. The second phrase is data processing. As we discussed before, user interactions on Facebook can be categorized into two groups: general interactions and emojis. According to the data provided by Facebook, there are three types of general interactions ("Likes", share, and comment), and five types of emojis ("Love", "Haha", "Wow", "Sad", and "Angry"). The number of these user interactions is public and accessible via the Facebook API. In the last phrase, we deployed BNN model to process multivariate input data and demonstrated that the prediction accuracy of color trends can be improved by including user interactions, especially emojis. We utilize social media features and combine with previous sales data as inputs of the model. Two baseline models were also implemented.

3.2　Bayesian Neural Networks

There are two reasons for us to apply BNN in this paper. The first one is that a careful probabilistic representation of uncertainty is needed. For prediction problem, limited knowledge is obtained, and it is impossible to exactly describe outcomes. Also, the lack of training data causes epistemic uncertainty. The second one is preventing overfitting. Many machine learning methods are data hungry. Much business-related data is scarce and makes the model tend to overfitting. Although lots of studies have applied machine learning methods to solve practical business problem, few of them notice the potential problem of overfitting. In this paper, we demonstrate the poor performance of standard neural networks and SVR on scarce business data and introduce BNN to solve this problem.

From a probabilistic perspective, standard NN training is a kind of maximum likelihood estimation (MLE) for the weights. Bayesian inference helps to generate a complete posterior distribution and provide parameters (e.g. w) a distribution instead of a single value. Therefore, BNN combine the strengths of stochastic modelling and standard neural networks. In this paper, BNN helps to estimate uncertainty of limited data, learn the model structure automatically, and avoid overfitting.

4　Data

This study uses a weekly sales dataset provided by an international clothing retailer. Sales data includes 31 styles and records transaction details of each style, such as color, sales quantity, sales amount, and average price. We obtained this data for the period between January 2015 and December 2016 (106 weeks).

We chose Facebook to get social media information. The most important reason is that Facebook allows users to react to posts by emojis. It provides us a valuable opportunity to study the operational value of emojis. Most companies have both Facebook and Twitter account to do marketing and always publish similar posts on these two platforms. Thus, we only need to choose one platform as the data source. Besides, Facebook offers public application programming interface (API) to access post information of public pages.

We obtained a collection of public posts from Facebook via the Facebook API. For each post, the API provided the text content, the created time, a unique post ID, the number of "Likes", the amount of interactions (e.g. shares and comments), and the amount of reactions (e.g. "Love", "Angry", "Haha", "Sad", and "Wow"). We predicted clothing sales using Facebook posts, which are all from one of the four types of Facebook public pages. They are brands, magazines, designers and influencers. We also set two criteria to choose qualified pages. First, "Likes" count of the page should be larger than 1,000. Second, the page must be authentic page for public figures, media companies and brands. In this way, we can get highly relevant posts and exclude useless information. Finally, we had 951 accounts in total. The data collection period for sales amounts is from January 2015 to December 2016, giving a total of 106 weekly sales amounts. Then we collect Facebook data recorded from December 2014 to December 2016 (totally 145,741 posts). The details of categorization of Facebook data are shown in Table 1.

Table 1. Facebook data source.

Category	No. of accounts	No. of posts
Brand	469	70,336
Magazine	93	15,207
Influencer	260	44,315
Designer	129	15,883
Total	951	145,741

Table 2 describes the social media features used in this paper. We constructed features on color level, since the purpose is predicting color trend. We observed the data carefully and decided not to consider "Angry" in this paper. Because we focus on posts related to fashion product, almost no one feel angry with these posts. All posts in our dataset have zero "Angry" emoji.

Table 2. Feature description.

Category	Item	Description
Post	Count	The number of selected posts mentioned certain color for one week
General interaction	Like	The total number of Like of selected posts
	Share	The total number of Share of selected posts
	Comment	The total number of Comment of selected posts
Emoji	Love	The total number of Love of selected posts
	Haha	The total number of Haha of selected posts
	Wow	The total number of Wow of selected posts
	Sad	The total number of Sad of selected posts

5 Evaluation

In this section, we report an experiment conducted to achieve our research purpose. To show the operational value of abundant social media data, we choose six different combination of inputs (Table 3). According to the retailer's suggestion, four-week (one month) ahead forecast is acceptable. Four prediction time windows (one to four weeks) were tested to achieve the best performance. We denote sales in week t in color m as S_{mt}, the number of total posts as $Count_{mt}$, the number of share as $Share_{mt}$, the number of "Likes" as $Like_{mt}$, the number of comments as $Comment_{mt}$, the number of "Love" as $Love_{mt}$, the number of "Wow" as Wow_{mt}, the number of "Haha" as $Haha_{mt}$, and the number of "Sad" as Sad_{mt}. To demonstrate the performance of the Bayesian Neural network, we evaluate it against popular machine classification methods, including support vector regression (SVR) and neural networks (NN). We choose a two-layer neural network with one hidden layer and one output layer as baseline model. Because the Bayesian neural network adopted in this paper has two layers.

Table 3. Input description.

Name	Input	Description
Input 1	(Sm(t-4))	One-week sales data
Input 2	(Sm(t-4), Countm(t-4), Sharem(t-4), Likem(t-4), Commentm(t-4))	One-week sales and general interaction data
Input 3	(Sm(t-4), Lovem(t-4), Wowm(t-4), Haham(t-4), Sadm(t-4))	One-week sales and emoji data
Input 4	(Sm(t-4), Countm(t-4), Sharem(t-4), Likem(t-4), Commentm(t-4), Lovem(t-4), Wowm(t-4), Haham(t-4), Sadm(t-4))	One-week sales, general interaction, and emoji data
Input 5	(Sm(t-4), Countm(t-4), Sharem(t-4), Likem(t-4), Commentm(t-4), Lovem(t-4), Wowm(t-4), Haham(t-4), Sadm(t-4), Sm(t-5), ..., Sadm(t-5))	Two-week sales, general interaction, and emoji data
Input 6	(Sm(t-4), Countm(t-4), Sharem(t-4), Likem(t-4), Commentm(t-4), Lovem(t-4), Wowm(t-4), Haham(t-4), Sadm(t-4), Sm(t-5), ..., Sadm(t-5), Sm(t-6), ..., Sadm(t-6))	Three-week sales, general interaction, and emoji data

The experiment results are presented in Table 4. The dataset is divided into training sets (80%) and testing sets (20%). All input data has been normalized. We use the root-mean-square error (RMSE) to evaluate models. 10-fold cross-validation is also used in evaluation. The results show that BNN outperforms all the other methods in terms of RMSE. Furthermore, the results demonstrate the BNN's ability of prevent overfitting. User interactions have significant impact on the performance of predictive models. In the following, we discuss the evaluation results in more detail.

Table 4. RMSE for different methods and variables.

Model data	BNN	NN	SVR
Input 1 (1 variable)	0.0533	0.0599	0.0736
Input 2 (5 variables)	0.0518	0.0525	0.0696
Input 3 (5 variables)	0.0498	0.0512	0.0644
Input 4 (9 variables)	0.0471	0.0510	0.0643
Input 5 (18 variables)	0.0425	0.0600	0.0647
Input 6 (27 variables)	0.0460	0.0702	0.0645

5.1   Bayesian Neural Networks

Considering the poor performance of SVR, we only compare BNN with NN in this part. As shown in Table 4, models with general interactions (Input 2) perform better than the ones without general interaction (Input 1). It is an expected result and coincides with previous studies. The Table 4 also shows that, adding emoji information can improve the performance. BNN and NN with Input 3 outperforms the same model with Input 1. In addition, the model with emoji data (Input 3) performs better than the model with general interaction data (Input 2). These results have significant practical implication. General interactions on social media have been regarded as valuable indicators in operational forecasting. Our results show that emojis are more helpful in color trend forecast. The value of a better color trend forecasting to fashion retailers can be substantial, since accurate forecast can help retailers to adjust the quantity of product in different colors, reduce inventory cost, and avoid stock out situation. We also test the model with both general interaction and emoji data (Input 4). The result shows that the combination of general interaction and emoji data can further improve the accuracy of forecasting. Every function on social media is well designed to satisfy user's different social demands. These functions help retailers and researchers understand potential customers from various aspects.

To further study the predictive power of each type of emojis, we conducted a follow-up experiment. The predictive model is Bayesian Neural network. Experiment setting is the same as the previous experiment. Forecast lead time L is four weeks. We denote sales in week t in color m as Smt, and other variables follow the same format. As Table 5 shown, the model with "Love" performs better than others.

Table 5. RMSE for different emojis.

Input	$Sm(t-4)$, $Lovem(t-4)$	$Sm(t-4)$, $Wowm(t-4)$	$Sm(t-4)$, $Haham(t-4)$	$Sm(t-4)$, $Sadm(t-4)$
RMSE	0.0485	0.0524	0.0516	0.0553

5.2 BNN vs NN vs SVR

The overall superior performance of BNN reveals the advantage of BNN on multivariate problem, especially when it comes to large dimensional data. For the two baseline models, the performance of SVR is relatively stable when the number of inputs increases. NN performs well using limited inputs. When the dimension of inputs becomes relatively high, the performance is poor because of overfitting. Compared to NN, BNN is much better when dealing with high-dimensional data. Although the RMSE increases for Input 5 and Input 6, BNN still works better than NN and SVR.

6 Conclusions and Future Work

In this paper, we conduct color trend prediction and improve the performance of the forecasting framework by two steps. The first one is that we incorporate emojis into our framework and demonstrate the predictive value of emoji data. The second one is that we apply Bayesian Neural Network to prevent overfitting problem that is caused by the increase of features.

Improving color trend forecasts accuracy can help companies in retail inventory management and thus lead to operational benefits. As an important element of fashion product, colors have significant impact on product sales. Color trend forecasts provide more fine-grained results compared to sales forecasts on product-level. At the same time, color trend forecasts need more market information to follow the trend. Many fashion retailers, especially "fast fashion" companies, require accurate forecasts to make response to the market quickly. Thus, accurate color trend forecasts are essential in the fashion industry. Our study shows that emojis on social media can significantly improve the accuracy of color trend forecasts. Besides novel variables, we also consider the multivariate data analysis and apply Bayesian Neural Network to avoid overfitting problem. Our work can inspire companies and researcher to apply various user interaction data in the forecasting model. We will continue to dig the operational value of emojis and apply other machine learning methods.

Acknowledgements. The authors thank to reviewers and colleagues for their constructive comments on the paper.

References

1. Moore, M., Fairhurst, A.: Marketing capabilities and firm performance in fashion retailing. J. Fashion Mark. Manag. Int. J. **7**(4), 386–397 (2003)
2. Yu, Y., Hui, C.L., Choi, T.M.: An empirical study of intelligent expert systems on forecasting of fashion color trend. Expert Syst. Appl. **39**(4), 4383–4389 (2012)
3. Kosinski, M., Stillwell, D., Graepel, T.: Private traits and attributes are predictable from digital records of human behavior. Proc. Nat. Acad. Sci. USA **110**, 5802–5805 (2013)
4. Chan, C.S.: Can style be measured? Des. Stud. **21**(3), 277–291 (2000)
5. Yu, Y., Choi, T.M., Hui, C.L., Ho, T.K.: A new and efficient intelligent collaboration scheme for fashion design. IEEE Trans. Syst. Man Cybern.-Part A Syst. Hum. **41**(3), 463–475 (2011)
6. Sun, Z.L., Choi, T.M., Au, K.F., Yu, Y.: Sales forecasting using extreme learning machine with applications in fashion retailing. Decis. Support Syst. **46**(1), 411–419 (2008)
7. Miller, H., Thebault-Spieker, J., Chang, S., Johnson, I., Terveen, L., Hecht, B.: "Blissfully happy" or "ready to fight": Varying Interpretations of Emoji. In: Proceedings of ICWSM (2016)
8. Eisner, B., Rocktäschel, T., Augenstein, I., Bošnjak, M. Riedel, S.: emoji2vec: Learning Emoji Representations from their Description. arXiv preprint arXiv:1609.08359 (2016)
9. Dresner, E., Herring, S.: Functions of the nonverbal in CMC: emoticons and illocutionary force. Commun. Theor. **20**(3), 249–268 (2010)
10. Vidal, L., Ares, G., Jaeger, S.R.: Use of emoticon and emoji in tweets for food-related emotional expression. Food Qual. Prefer. **49**, 119–128 (2016)
11. Winters, P.R.: Forecasting sales by exponentially weighted moving averages. Manag. Sci. **6**(3), 324–342 (1960)
12. Alon, I., Qi, M., Sadowski, R.J.: Forecasting aggregate retail sales: a comparison of artificial neural networks and traditional methods. J. Retail. Consum. Ser. **8**(3), 147–156 (2001)
13. Choi, T.M., Hui, C.L., Ng, S.F., Yu, Y.: Color trend forecasting of fashionable products with very few historical data. IEEE Trans. Syst. Man Cybern. Part C (Appl. Rev.) **42**(6), 1003–1010 (2012)
14. Li, W., Chau, M.: The predictive power of online user engagement on product sales. In: Proceedings of the Workshop on E-Business (WEB 2017), Seoul, South Korea (2017)
15. Frank, C., Garg, A., Sztandera, L., Raheja, A.: Forecasting women's apparel sales using mathematical modeling. Int. J. Clothing Sci. Technol. **15**(2), 107–125 (2003)

16. Cui, D., Curry, D.: Prediction in marketing using the support vector machine. Mark. Sci. **24** (4), 595–615 (2005)
17. West, P.M., Brockett, P.L., Golden, L.L.: A comparative analysis of neural networks and statistical methods for predicting consumer choice. Mark. Sci. **16**(4), 370–391 (1997)
18. Denker, J., et al.: Large automatic learning, rule extraction, and generalization. Complex Syst. **1**(5), 877–922 (1987)
19. Hernández-Lobato, J.M. Adams, R.: Probabilistic backpropagation for scalable learning of bayesian neural networks. In: International Conference on Machine Learning, pp. 1861–1869 (2015)
20. Xiong, H.Y., Barash, Y., Frey, B.J.: Bayesian prediction of tissue-regulated splicing using RNA sequence and cellular context. Bioinformatics **27**(18), 2554–2562 (2011)
21. Utama, R., Piekarewicz, J., Prosper, H.B.: Nuclear mass predictions for the crustal composition of neutron stars: a Bayesian neural network approach. Phys. Rev. C **93**, 014311 (2016)

Artificial Intelligence (AI) and Cognitive Apportionment for Service Flexibility

Xue Ning[1](✉), Jiban Khuntia[1], Abhishek Kathuria[2],
and Benn R. Konsynski[3]

[1] University of Colorado Denver, Denver, CO 80204, USA
Xue.Ning@UCDenver.edu
[2] University of Hong Kong, Pokfulam, Hong Kong
[3] Emory University, Atlanta, GA 30322, USA

Abstract. Artificial Intelligence (AI) is evolving from a technology to an enabler of service processes and delivery. With the increasing importance of AI as a service interface, integrating and aligning AI technology to effective service delivery is a critical challenge. In this study, we theorize *cognitive apportionment*, consisting of AI-human service synchronization and AI-system service consistency attributes, as a dynamic capability that could address issues arising from shifting decision processes. We argue that integrating and balancing human-AI-system dialogue mechanisms while rendering services is a key to leverage and achieve desired service outcomes. We develop a set of propositions encompassing the mediating effect of cognitive apportionment and the moderating role of AI-enabled service portfolio to influence service flexibility. Preliminary analysis with 80 respondents supports our propositions. This research offers new insights into the underlying mechanisms of how AI can create value.

Keywords: Artificial intelligence · Cognitive apportionment ·
Service flexibility

1 Introduction

Artificial intelligence (AI) is emerging to be central to businesses. AI is providing rich, personal, and convenient services across industries, including retail, utilities, manufacturing, healthcare, and education. In 2016, companies invested $26–39 billion in artificial intelligence (Chui and Francisco 2017). Accenture estimates that business revenues from AI investments will increase by 38% between 2018 and 2020 (Shook and Knickrehm 2018). Use of AI for applications is expected to change how organizations utilize, manage, and deliver services to their customer—either in a brick-and-mortar service setting or a pure-play digital service setting enabled by the internet. Amazon's Alexa and Google's Home in hotel rooms are examples of service offerings through AI to enhance the user experience, and thus, create value for companies.

Preparedness for AI-enabled competitive advantage is not only dependent on the choice of the technology but also on the appropriate use of the technology for business purposes (Davern and Kauffman 2000). The challenge is also leveraging the existing data, information, processes, workflows, and the extent to which the AI technology can

© Springer Nature Switzerland AG 2019
J. J. Xu et al. (Eds.): WEB 2018, LNBIP 357, pp. 182–189, 2019.
https://doi.org/10.1007/978-3-030-22784-5_18

orient these elements to meet business objectives such as providing efficient and effective services. Big companies such as Amazon and Google are competing over AI devices and pushing it to be a central part of many applications and services, however, just installing the Echo or Alexa devices is not enough. Rather, orienting these devices and digital assistants to interface with both existing human-delivered service and information technology (IT) enabled service deliveries are highly critical for successful AI-enabled service provision. Thus, while the human-support based decisions for services are changing in recent times; information and analytics, along with the data and AI supported models are facilitating machine enabled cognitive support for services (Karhade et al. 2015; Watson 2017).

Cognitive apportionment – a concept proposed by Konsynski and Sviokla (1993) – suggests that cognitive responsibilities can be allocated to human or system in the back end enabling process. With the advancement of technologies, human-system and system-system dialogues are replacing the erstwhile purely human-human dialogues. Value creation will then depend on the allocation of decision rights not only to humans but also to systems. The challenge is higher for service-enabling businesses because AI can be embedded fully in a production or manufacturing process to make an automated system (Sidorova 2018), while services would always demand a set of person-orientation and user-centricity enabled processes (Peters and Zaki 2018). As usual, service delivery and related IT enablement of services have been facing the challenge of balancing act across high-tech and high-touch nature of services. In other words, while technology has enabled customization and flexibility in a great way for services, it also has taken away the 'spontaneous delight' derived from the service experience (Bitner et al. 2000). Thus, it is imperative to have a balanced AI-human-system dialogue mechanism while rendering services. Responsible cognitive *reapportionment* of human-human dialogues to the human-system and system-system dialogues is a need of service process, specifically in the infusion of AI in services with effectiveness and customer satisfaction focus.

To manage apportionment complexity arising from the change from the eras of human-human interactions to the human-system and system-system interactions, there would be a transition state involving a set or combinations of human-system interactions. Arguably, the current times are facing this threshold of the interaction-transition stage and providing a set of challenges to the business environment. It is imperative that managers should pay more attention to these developments that include the decision rights allocations through the artificial and cognitive automation processes (Scheepers et al. 2018). However, how the firms get prepared for the balancing act of human-system and system-system cognitive reapportionment complexity, what is the role of AI in the cognitive apportionment, and what is the influence on service flexibility, are not clear.

The main research question of this study is how cognitive apportionment enables AI technology for service flexibility as a mediating factor. We conceptualize the new construct cognitive apportionment as reflected through two facets: AI-human service synchronization and AI-system service consistency. Service flexibility is defined as a second order reflective construct consisting of service responsiveness and service scalability. We also consider the moderating effect of the AI-enabled service portfolio of a firm on the relationship between AI technology and cognitive apportionment. We draw on the dynamic capability hierarchy concept (Madhavaram and Hunt 2008; Winter 2003), and argue that, to apportion value from AI technologies a firm needs to effectively deploy and leverage different levels of resources for developing service models so it can reliably achieve desired outcomes. We develop the propositions based on the theoretical foundations for the dynamic capability hierarchy and apply to the AI technology-cognitive apportionment-value creation framework. We offer four propositions for this approach to AI-value creation. Preliminary results with 80 firms based in India, that leverage artificial intelligence computing for service rendering, suggest broad support for the articulated concepts and propositions. This study provides a new theoretical understanding of cognitive apportionment applied to the AI-enabled service context. We offer new managerial and theoretical insights.

2 Prior Research and Theoretical Background

The goal of AI technology is to create machines that exhibit human-like intelligence (Watson 2017). AI development started in the 1950s (Moreira Nascimento et al. 2018; Sidorova 2018), and because of high-performance computing architecture, advanced deep neural networks, and unprecedented data richness, AI is emerging as a set of technologies with influence across all industries. AI technology has progressed to include construction of computer programs that display intelligence similar to human intelligence and complement human intelligence (Moreira Nascimento et al. 2018). Konsynski and Sviokla (1993) articulate the cognitive apportionment concept, which has the core idea that cognitive responsibilities can be allocated to human or systems. Especially, with the technology development and changing business environment, it is important for firms to allocate decision rights not only to the human but also to the systems, in order to create IT business values. Because of the rising of AI technology, researchers expect the upcoming cognitive generation of decision making and companies such as IBM are developing cognitive computing (Watson 2017). The previous literature has demonstrated the case of using AI-enabled cognitive automation to handle customer inquiries (Scheepers et al. 2018). From a case of decision making in ambiguous procurement tasks, researchers showed the role of software and human agents in adaptive decision making (Nissen and Sengupta 2006). In this paper, we study cognitive apportionment from both the human and system perspectives and explore the role of AI in balancing human and system cognitive responsibilities in the service delivery process.

Existing research suggests that AI-enabled service rendering might help companies to leverage the AI technology to "manage and (re-) configure their customers' journeys in a flexible manner" – highlighting the important role that AI technology plays to orchestrate service flexibility (Peters and Zaki 2018). Service flexibility refers to the ability of a firm to reallocate and reconfigure its organizational resources, processes, and strategies upon environmental changes to offer high-quality services in this study. We capture service responsiveness and service scalability to reflect the level of service flexibility (Tiwana and Konsynski 2010). Also, an organization may provide multiple types of services to its customers, and an efficient service portfolio will be the key in this process. In this paper, AI-enabled service portfolio refers to the ability of a firm to exploit AI technology to support and enhance its service portfolio. There are five elements in the AI-enabled service portfolio, i.e., design, development, deployment, management, and maintenance of the service portfolio using AI technology (Kathuria et al. 2018b). According to the resource-based view, in order to gain competitive advantage and create business value, companies are trying to improve their customers' experience in the service offering process (Peters and Zaki 2018).

Research is sparse to provide empirical support as well as identify the enablers of the AI technology and service flexibility process. This study tries to address the research gap by providing a cognitive apportionment mediating, and an AI-enabled service portfolio moderating mechanism.

We anchor to the dynamic capability hierarchy concept that suggests a firm creates value in its operational state using stationary processes from zero-level capabilities (Winter 2003). Higher order capabilities create, extend, or modify the lower-order capabilities by governing the rate of change of ordinary capabilities to create value (Collis 1994). This hierarchical approach needs to be developed, coordinated, and seamlessly integrated to provide services (Madhavaram and Hunt 2008). Based on this theoretical background, we propose that the AI technology influences service flexibility through cognitive apportionment mediating effect. Furthermore, we posit that AI-enabled service portfolio is the moderator of the relationship between AI technology and cognitive apportionment. We offer four propositions corresponding to these relationships.

3   Theoretical Propositions

We argue that AI technology realizes cognitive apportionment in the service delivery context. As mentioned earlier in this paper, cognitive apportionment realized through two ways: (1) the synchronization between AI and human-provided services and (2) the consistency between AI and other IT systems enabled services. First, AI-based performance improvement through efficient and effective perception, prediction, prescription, and integration as an ability that helps managers to allocate the cognitive responsibility between human and systems based on the characteristics of different types of services (Barton et al. 2017). Second, AI augments human workers' cognitive skills and creativity and also enhances systems' flexibility, speed, and scale through cognitive computing (Wilson and Daugherty 2018). Managers can seek the balance and optimize the resources allocation through cognitive apportionment or reapportionment. The extent to which a firm can balance the allocation of decision support or cognitive responsibilities between human and systems for high-quality service seems to be dependent upon a firm's ability to deploy, operate, and functionalize AI technology. Thus, we propose that:

Proposition P1: AI Technology is positively associated with Cognitive Apportionment.

Cognitive apportionment of a firm in the service delivery process is about the firm to synchronize the AI technology and human-delivered service as well as make the AI technology, and other IT systems enabled services to be consistent. Firms with better cognitive apportionment can improve the responsiveness and scalability when they deliver services. First, synchronization between AI technology and human-provided service improves the service responsiveness (Shook and Knickrehm 2018). AI and human intelligence joined forces to respond to customer inquiries in a swift and human-oriented manner. This synchronization also enhances the service scalability. For certain repetitive activity in the service offering process, AI can take over and handle the service provision in large scale with its powerful computing ability (Wilson and Daugherty 2018). Second, consistency between AI technology and other IT systems enabled service could increase both the service responsiveness and service scalability. On the one hand, the consistency improves the fluency of the workflow and thus achieves higher speed and better productivity (Wilson and Daugherty 2018). On the other hand, the consistency facilitates information management and sharing in different sectors within a firm and thus ensures the scalability of service provision. Accordingly, we propose that:

Proposition P2: Cognitive Apportionment is positively associated with Service Flexibility.

AI technology can affect a firm's service flexibility through the mediating role of cognitive apportionment. The higher ability of a firm to deploy, operate, and functionalize AI technology could enable the firm to improve the cognitive apportionment or reapportionment performance in a manner that increase the synchronization of AI technology and human-provided services, as well as the consistency between the AI technology and the other IT systems enabled services. In turn, this improved cognitive apportionment performance leads to higher service flexibility, including quicker service responsiveness and larger service scale. In contrast, without cognitive apportionment, the firm may not achieve superior service flexibility just with AI technology. Building on this logic, we expect the mediating role of cognitive apportionment between the relationship of AI technology and service flexibility, and we formulate the proposition:

Proposition P3: Cognitive Apportionment mediates the relationship between AI Technology and Service Flexibility.

If a firm wants to achieve high-level cognitive apportionment in the service delivery process, a proper match between AI technology and the firm's internal service portfolio is important. First, a well-organized and managed firm-level service portfolio by using the AI technology enables the firm to respond to the market shifting. Firms that apply AI to design and deploy service portfolio could sense and adapt to the change and thus adjust the allocation of resources between human-provided services and IT systems enabled services. Second, with the same AI technology, if a firm has higher ability to develop and maintain its service portfolio, it is easier and clearer for the firm to synchronize the AI technology and human delivered services as well as to keep the consistency between AI technology and other IT systems enabled service offering. Based on this rationale, we propose that:

Proposition P4: AI-enabled Service Portfolio positively moderates the relationship between AI Technology and Cognitive Apportionment.

4   Empirical Validation

Our theoretical concept and propositions empirically validated through a primary study. We next provide a roadmap for methods, including an overview of a preliminary dataset, variables, the rationale for using partial least squares (PLS) and other techniques to support our ability to make inferences, and the benefits they offer for learning about cognitive apportionment and service flexibility. A working paper, with the authors, is available for more details about the methodology followed. A brief about the data collection and analysis is given below, and details are omitted for brevity.

4.1   Data Collection

Our propositions can be validated through a cross-sectional matched-pair field survey. We have conducted such a survey of organizations in India, an emerging economy with a large number of users and suppliers for AI-based services. To minimize confounding factors due to uneven economic development in India, we have developed a sample with data for firms that are located near two commercial hubs in western and southern India (Kathuria et al. 2018a).

Pre-tested and back-translated survey instruments were pilot tested with a small sample from the targeted population for reliability, convergent and discriminant validity, and predictability. Matched-pair data were collected through anonymous surveys of volunteering organizations administered using a dual online-offline mode (Kathuria et al. 2018b).

4.2   Partial Least Square Analysis

We used partial least squares (PLS), a structural equation modeling (SEM) technique, to empirically validate the propositions. PLS estimates interrelated dependence relationships and handles second-order formative constructs better than covariance-based SEM. It assesses a measurement model for the conceptual framework and makes no assumptions about data normality. We began with an assessment of the measurement model and its factors, followed by a structural model assessment. Our preliminary path analysis and mediation results with 80 responses offer evidence for a path for firms to follow for extracting strategic value from AI technology.

5   Discussion

Our objective was to propose and validate that AI technology for firm value creation, using cognitive apportionment as a mediating factor is a feasible theoretical approach. This mediation is important to establish the human-system coordination synchronization and system-system integration consistency in the back end, which in turn enables the AI technology interface to deliver effective services. Through the existing AI and IT capability literature, and the dynamic capability hierarchy perspectives, we identified the key concepts to explain AI-cognitive apportionment-service flexibility process.

We aim to provide a nuanced understanding of the role of cognitive apportionment to enable AI to increase service flexibility.

This study offers several contributions to IS research: it adds to the sparse AI business value literature, contributes to the early discussions around capability hierarchy by applying it to AI context, and suggests that there is a way the system-human and system-system combination can bring the higher value of AI technology implementation for value creation. Our study also has practical implications. For example, regarding the widespread debate evokes around AI that AI is going to replace human workers, our study argues that AI and human workers could join forces to create greater value.

References

Barton, D., Woetzel, J., Seong, J., Tian, Q.: Artificial Intelligence: Implications for China, pp. 1–20. McKinsey Global Institute, New York (2017)

Bitner, M.J., Brown, S.W., Meuter, M.L.: Technology infusion in service encounters. J. Acad. Mark. Sci. **28**(1), 138–149 (2000)

Chui, M., Francisco, S.: Artificial Intelligence the Next Digital Frontier?, pp. 1–75. McKinsey Global Institute, New York (2017)

Collis, D.J.: Research note: how valuable are organizational capabilities? Strateg. Manag. J. **15**(S1), 143–152 (1994)

Davern, M.J., Kauffman, R.J.: Discovering potential and realizing value from information technology investments. J. Manag. Inf. Syst. **16**(4), 121–143 (2000)

Karhade, P., Shaw, M.J., Subramanyam, R.: Patterns in information systems portfolio prioritization: evidence from decision tree induction. MIS Q. **39**(2), 413–433 (2015)

Kathuria, R., Kathuria, N.N., Kathuria, A.: Mutually supportive or trade-offs: an analysis of competitive priorities in the emerging economy of India. J. High Technol. Manag. Res. **29**(2), 227–236 (2018a)

Kathuria, A., Mann, A., Khuntia, J., Saldanha, T., Kauffman, R.J.: Understanding the strategic value appropriation path for cloud computing in the organization. J. Manag. Inf. Syst. **35**(3), 740–775 (2018b)

Konsynski, B.R., Sviokla, J.J.: Cognitive Reapportionment: Rethinking the Location of Judgement in Managerial Decision Making. Working paper (1993)

Madhavaram, S., Hunt, S.D.: The service-dominant logic and a hierarchy of operant resources: developing masterful operant resources and implications for marketing strategy. J. Acad. Mark. Sci. **36**(1), 67–82 (2008)

Nascimento, A.M., da Cunha, V.C., Alexandra, M., de Souza Meirelles, F., Scornavacca, E., de Melo, V.V.: A literature analysis of research on artificial intelligence in Management Information System (MIS). In: Twenty-Fourth Americas Conference on Information Systems, New Orleans (2018)

Nissen, M.E., Sengupta, K.: Incorporating software agents into supply chains: experimental investigation with a procurement task. MIS Q. **30**(1), 145–166 (2006)

Peters, C., Zaki, M.: Modular Service Structures for the Successful Design of Flexible Customer Journeys for AI Services and Business Models–Orchestration and Interplay of Services. Working paper (2018)

Scheepers, R., Lacity, M.C., Willcocks, L.P.: Cognitive automation as part of Deakin University's digital strategy. MIS Q. Executive **17**(2), 89–107 (2018)

Shook, E., Knickrehm, M.: Reworking the Revolution, pp. 1–44. Accenture Strategy (2018). https://www.accenture.com/_acnmedia/PDF-69/Accenture-Reworking-the-Revolution-Jan-2018-POV.pdf

Sidorova, A.: Interests and agency in AI: the case of image with Inception 3 model. In: Twenty-Fourth Americas Conference on Information Systems, New Orleans (2018)

Tiwana, A., Konsynski, B.: Complementarities between organizational IT architecture and governance structure. Inf. Syst. Res. **21**(2), 288–304 (2010)

Watson, H.J.: Preparing for the cognitive generation of decision support. MIS Q. Executive **16**(3), 153–169 (2017)

Wilson, J., Daugherty, P.R.: Collaborative intelligence: humans and AI are joining forces. Harvard Bus. Rev. **96**(4), 115–123 (2018)

Winter, S.G.: Understanding dynamic capabilities. Strateg. Manag. J. **24**(10), 991–995 (2003)

How Long Will Your Videos Remain Popular? Empirical Study of the Impact of Video Features on YouTube Trending Using Deep Learning Methodologies

Min Gyeong Choe, Jae Hong Park[(✉)], and Dong Won Seo

Kyung Hee University, Dongdaemun-gu, Seoul 08544, South Korea
{alsrud6949, jaehp, dwseo}@khu.ac.kr

Abstract. YouTube has become one of the most influential channels in recent years. There are an enormous number of videos on the platform, but few of them are popular, getting placed in the "Trending" section. But, videos on this list have different stories. Some of them will get constant popularity and others will fade out. Many researchers have analyzed what will make a video become popular. However, no study has focused on how long a video maintains its popularity. In addition, the content similarity between the thumbnail image and the title has been neglected, although it appears to play an important role in social media posts (e.g. blogs, Instagram). We measure the variable, content similarity, by analyzing the thumbnail image and text. This study investigates the impact of this new variable on popular videos' survival to give YouTubers and advertisers insights into video marketing. Also, our suggested approach can achieve new academic results in the research of YouTube.

Keywords: YouTube trending · Video popularity · Content similarity · Deep learning method

1 Introduction

In 2012, the music video "Gangnam style" made a stir and became the first video to hit 1 billion views on YouTube. Later, it even hit 2 billion views in 2014. In addition, many people have generated a lot of User Generated Contents (UGC) responding to "Gangnam style". Accordingly, this video has become a legend in online video history, bringing Psy, the artist, a big success. "Gangnam style" has attracted much attention and maintained its popularity for a couple of years. Psy won $4.6 million in profits with this single video [6]. However, we still don't know what made "Gangnam style" popular for so long. Why this happen? In this paper, we investigate what makes videos continue to be popular. That is, this study examines popular videos' common features obtained by users and by creator (e.g. views, likes, dislikes, comments, thumbnail, title) to see what makes videos popular for a long time.

"Gangnam style" was interesting enough to make many researchers and marketers examine the video's popularity and virality on YouTube [6]. Many researchers have focused on identifying the factors that make videos popular or viral [2–4, 6, 12].

© Springer Nature Switzerland AG 2019
J. J. Xu et al. (Eds.): WEB 2018, LNBIP 357, pp. 190–197, 2019.
https://doi.org/10.1007/978-3-030-22784-5_19

At first, most of them investigated the roles of view counts or view's growth patterns on popularity [2, 3, 12], while others analyzed videos' popularity with time-series approaches [2, 9]. Recent studies have examined sentiments of image and text to predict the popularity of videos [3, 7]. Still, it has not yet been considered how consistently popular the already popular video would be. Thus, this study investigates what makes popular videos be popular continuously with thumbnail image and text.

Nowadays, news about Trending videos explode with titles like "10 Top Trending videos of this year". However, we notice that only 10 or 20 videos attract public attention among the Daily 200 videos at YouTube over the course of a year. This indicates that we need to see about how long and which trending videos get attention and win popularity. To examine this question, we focus on what attracts consumers to click the video. This basically relies on the thumbnail image and title, which can be shown before clicking the video. If the thumbnail contains texts, these can make the video credible especially when matching the title. Content similarity indicates the similarity in terms of topics from image and text [10]. This variable can indicate the credibility of the video, influencing on video popularity. While a few studies have considered thumbnails as a factor resulting in popularity [3, 4], what has been lacking is to examine the impact of content similarity of thumbnail and title on video popularity. But this thumbnail feature is so important enough to be emphasized among YouTube creator experts. Even, the creator academy supported by YouTube has a class like "make effective thumbnails and titles". This supports why we need to analyze these factors, thumbnail and title, in research about YouTube videos.

Some popular videos on YouTube have seen great success, and the platform itself, YouTube, has also become more influential as a media outlet. Accordingly, "creators", who make that powerful videos on YouTube gain fame and riches. Usually they can earn profits from advertisements set upon registering their own videos or making video advertisements requested by companies or brands. The more they create popular videos, the better they make a gain from them. So, for creators, making their videos attractive could be important to gain more profits and popularity. For example, Felix Kjellberg, one of the most influential creators on YouTube, has 40,315,481 subscribers and earn $1 million–$16.5 million as a yearly income from YouTube [5]. In addition, marketers know that it is so important to find creators who will promote their brand and stay popular for longer. In the case of videos made by a company itself, popular videos can have advertising earnings and additional profits derived from improving brand images and directly increasing the purchase of products. That is why we need to figure out what makes video stay popular for a long time and determine how long it will be popular.

According to YouTube, "Trending" is a section that ranks top popular videos based on view count, the rate of growth in views, where views are coming from, and the age of the video. But, not all the videos on the Trending list can survive long to have a great success. That is, depending on the features of the trending video (e.g. thumbnails, views, etc.), some trending videos will stay on the list for a long time and others will not. However, little is known about how such features influence the lifetime of videos at the Trending section of YouTube.

Regarding images and text-related studies, some have shown that images and text are important for attracting users to blogs or social media posts. In the case of blogs, since the images and texts are presented simultaneously on posts, users will intuitively recognize them as a single content. Therefore, more coherent content will anchor people more and will get them to engage actively in the post [10]. Having coherent content also gives the source credibility on social media and brings more favorable user attitudes toward it [1]. In this regard, if the thumbnail and title of an online video have high content similarity, then users will tend to watch them more. Thus, the main goal of this study is to investigate that what thumbnail features contain can make videos remain popular in the Trending section of YouTube by examining the content similarity between the thumbnail (image) and title (text).

To answer this research question empirically, we used data from the YouTube Trending video list in the US for one month (May 2018). The data includes views, likes, dislikes, comments, title, channel title, upload time, thumbnail, description, and others. Using text mining and image mining, we measure the content similarity between title and thumbnail text to analyze whether this attract consumers and make video keep popular. We will also investigate that view count and other indicators make a video survive longer in the Trending list, using the Cox regression model.

Our research can give a practical guide to both YouTube content creators and marketers by demonstrating how the thumbnail image and text can make people engage more and to help the video survive longer on the Trending list. Creators can use our findings to make their popular videos gain even more popularity for a longer period. Also, marketers can utilize effective strategies for video advertising through YouTube. Academically, this study can contribute to the e-commerce literature as we empirically analyze log data from YouTube trending using machine learning, deep learning methods, and the econometric model.

2   Data

This paper selected YouTube as the context of study. Through YouTube, users can make their own videos, watch videos, and communicate with each other through videos or post their content on other social media platforms to share the videos with their friends [11]. Many marketers consider YouTube a remarkable tool to advertise their brands. First, social video marketing like YouTube can attract more people than traditional commercials. For example, in the entertainment industry, people's interest is causing a shift from sound sources and record albums to multimedia like music videos. Also, people prefer short videos on YouTube to watch broadcast news. An industrial report suggests that video traffic accounted for 73% of all customer internet traffic in 2016 and will increase to 82% by 2021 (Cisco Study, 2017). This implies that the video market will be even bigger in the future. Second, many people have created and watched UGCs more and more, letting users feel more friendly connection with creators and the brands they use. For instance, the beauty industry pays attention to ads through YouTubers because customers can relate themselves to creators more easily than they do with celebrities.

One big feature of YouTube is that the videos can be shared freely with any user. Due to this feature, popular videos have a big viral or WOM effect. Advertising on YouTube does not end on the platform, but rather the content spreads widely. So, users (watchers) can be on the channel itself, attracting new consumers to the video on this platform. If videos get popular and consistently maintain that popularity, then the advertising effect would be enhanced. We therefore chose to study popular videos and the duration for which they are popular.

We use YouTube's Trending data, which has been collected daily in the USA as obtained by Kaggle. The daily top 200 videos are listed and have the number of views, likes, dislikes, and comments. Also, each video's thumbnail image and title that users can observe even before they watch videos are included in the data. We collected the top 200 videos daily for a month, resulting in 6,198 observations with missing 2 observation. These are all in the "News & Politics" category. We define our variables in Table 1 and present their descriptive statistics in Table 2. For survival analysis, we need the event variable and time variable as the dependent variables. To make the event variable, we first coded an Exit variable as follows.

$$Exit_{it} = \begin{cases} 1, & \text{if video } i \text{ exits at time } t \\ 0 & \end{cases} \tag{1}$$

The Exit variable indicates if the video exited from the Trending list at time t. The time variable is the duration of survival on the Trending list before exit. In the dataset,

Table 1. Description of variables.

Variables	Example
Exit	Binary indicator for the day of exit from the Trending list, set to 1 for exit
Duration	The length of time until exit
Ln_views	The number of accumulated views of the video at the trending date (logged)
Ln_likes	The number of likes video got up to the trending date (logged)
Ln_dislikes	The number of dislikes video got up to the trending date (logged)
Ln_comment	The number of comments videos got up to the trending date (logged)
Like ratio	The ratio of likes to the sum of likes and dislikes
Content similarity	The extent of content similarity between thumbnail and title

Table 2. Descriptive statistics of variables.

Variables	Obs	Mean	Std.dev	Min	Max
Exit	6,198	0.1182449	0.3229239	0	1
Duration	6,198	13.49718	5.830035	1	26
Ln_views	6,198	14.332	1.273445	10.93439	19.19886
Ln_likes	6,198	10.62826	1.474096	5.894403	15.53742
Ln_dislikes	6,198	7.320695	1.488168	2.833213	12.72326
Ln_comment	6,198	8.209749	1.800217	0	14.01872
Like ratio	6,198	0.9458607	0.0661844	0.3615977	0.9977
Content similarity	6,198	0.5838974	0.2008918	0	1

there are the number of daily views, likes, dislikes, and comments, which are time-variant variables. Also, each video has its own thumbnail image and title and we used these to make the content similarity variable.

3 Method

In this paper, we aim to examine what makes a video popular for a longer time. Thus, we used the Cox regression model, which is a survival model. We first assumed that videos watched by more people will survive longer in the Trending list. And if a video got more comments, then it will not exit from the list for longer. Unfortunately, other features such as ln_likes and ln_dislikes cannot be used due to the high correlation with views although like ratio measured by likes and dislikes has low correlation with views (Table 3). To test our research questions, we first estimated a time-dependent Cox regression model.

Table 3. Correlation of key variables.

Variables	[1]	[2]	[3]	[4]	[5]	[6]	[7]	[8]
[1] Exit	1.000	–	–	–	–	–	–	–
[2] Duration	−0.249	1.000	–	–	–	–	–	–
[3] Ln_views	0.001	0.285	1.000	–	–	–	–	–
[4] Ln_likes	−0.027	0.274	0.836	1.000	–	–	–	–
[5] Ln_dislikes	−0.013	0.261	0.843	0.777	1.000	–	–	–
[6] Ln_comment	−0.022	0.231	0.614	0.730	0.667	1.000	–	–
[7] Like ratio	−0.022	0.042	−0.003	0.266	-0.297	0.066	1.000	–
[8] Content similarity	0.018	−0.086	−0.167	−0.157	−0.133	−0.110	0.000	1.000

We also examined whether the time-invariant factor, content similarity, affects the length of video survival. If a video's thumbnail image and title have similarity in terms of topic, then it will stay on the Trending list longer. Due to an anchoring effect, people may watch the video with a higher similarity between the thumbnail and title. This similarity can be measured by both text mining and optical character recognition (OCR) analysis. First, we extract the text from thumbnail image using OCR model. Next, the text on both thumbnail and title will be analyzed using sentence2vec model, being represented as a vector. Then, we calculate the distance between two text vectors. That is, we can measure the content similarity between the text on thumbnail image and title, calculating the cosine similarity [10]. If the words match the title, it can be expected to be more credible and attract consumers' attention.

3.1 Variable Construction

To estimate the content similarity variable, we checked how similar the topic, image, and text are to each other. First, the OCR model we employed detects text objects in the images and extracted semantic features. Using the features, we gathered text on the thumbnail image, which can be the topic presented by a creator. Next, we separate the title sentence by word, make the list of word vectors and perform sentence2vec(sif), representing one sentence as one vector. Also, in terms of text from OCR model, we make one vector and calculate the distance between two vectors (title, thumbnail). Using cosine similarity, content similarity between the thumbnail and title were measured [10]. Cosine similarity was calculated by:

$$ContentSimilarity = \frac{c_{image}^T \cdot c_{text}}{\left\| c_{image} \right\| \cdot \left\| c_{text} \right\|} , c : \text{topic distributions} \tag{2}$$

3.2   Survival Analysis

A survival analysis is a statistical method for analyzing data on the occurrence of events. We use a Cox regression model for estimation. An outcome variable is the hazard rate of the video, which is the probability that the duration will end after time t, given that it has lasted until time t. In this study, since independent variables contain time-variant covariates, we employ the time-variant Cox regression model. The ln_views, like ratio and ln_comment are time-dependent, while the content similarity is independent of time. So, the variables at time t are denoted as:

$$X(t) = \{ln\_views_{it}, likeratio_{it}, ln\_comment_{it}, contentsimilarity_i\}$$

What we want to know is how much the probability of exit from the list increases for every 1-unit increase in each predictor, which is the hazard rate given the variables.

$$\lambda(t|X(t)) = \lambda_0(t)e^{\beta X(t)} = \lambda_0(t)e^{\beta_1 ln\_views_{it} + \beta_2 likeratio_{it} + \beta_3 ln\_comment_{it} + \beta_4 contentsimilarity_i} \tag{3}$$

Coefficients β_1, β_2, β_3, and β_4 are estimated by maximizing the log-partial likelihood function below. The estimated coefficients indicate how much each predictor influences the hazard ratio.

$$log\frac{\lambda(t|X(t))}{\lambda_0(t)} = \beta_1 views_{it} + \beta_2 likeratio_{it} + \beta_3 ln\_comment_{it} + \beta_4 contentsimilarity_i \tag{4}$$

4   Results

According to Table 4, content similarity affected hazard ratio of video negatively. If hazard ratios are less than 1, a unit increase in independent variable decreases the likelihood of exit given period by (1-hazard ratio) [8]. Therefore, the negative coefficient of content similarity means that a unit increase in it decreases the likelihood the video exit from the Trending list during the given days by about 79%. This can be interpreted that if the text on thumbnail match the title, the video has high probability to keep popular. On the other hand, other variables, ln_views and ln_comment, did not have any influence on the survival probability of video. But like ratio increased the likelihood of video survival significantly. It indicates that if the video earns more likes compared to dislikes, it tends to stay in the list for longer time.

Table 4. Result of cox model.

	Haz. Ratio	Std. err.	z	P > |z|	[95% conf. interval]	
Time-independent						
Content similarity	.2078	.1683	−1.94	0.052	.042475	1.016627
Time-dependent						
Ln_views	1.5837	.6714	1.08	0.278	.689901	3.635249
Ln_comment	.9528	.1326	−0.35	0.728	.725343	1.251611
Like ratio	.0000	.0000	−2.36	0.018	.000000	.117451

5   Future Plan

We collected videos in the "News & Politics" category only. So, we keep collecting other category videos for the generality. And once video catch the consumers' eyes, the next process will be to watch the video. In this process, the video characteristics, especially the contents are crucial. To get the video contents, we are planning to obtain the speech from the video, using web speech api. Therefore, our second research question is whether the coincidence between the expected and real contents of video will make video keep popular for longer time, eliciting more both attention and engagement. We hope that we can present further findings at the conference.

References

1. Argyris, Y.E.A., Xu, J.D.: Influencer marketing for increasing consumer engagement and brand connection. Commun. Assoc. Inf. Syst. **34**, 555–585 (2018)
2. Figueiredo, F., Benevenuto, F., Almeida, J.M.: The tube over time: characterizing popularity growth of YouTube videos. In: Proceedings of the fourth ACM International Conference on Web Search and Data Mining, pp. 745–754 (2011)
3. Fontanini, G., Bertini, M., Del Bimbo, A.: Web video popularity prediction using sentiment and content visual features. In: Proceedings of the 2016 ACM on International Conference on Multimedia Retrieval, pp. 289–292 (2016)
4. Fu, W.W.: Selecting online videos from graphics, text, and view counts: the moderation of popularity bandwagons. J. Comput. Mediated Commun. **18**(1), 46–61 (2012)
5. Holland, M.: How YouTube developed into a successful platform for user-generated content. Elon J. Undergrad. Res. Commun., 7(1) (2016)
6. Jiang, L., Miao, Y., Yang, Y., Lan, Z., Hauptmann, A.G.: Viral video style: a closer look at viral videos on YouTube. In: Proceedings of ACM International Conference on Multimedia Retrieval, p. 193 (2014)

7. Joglekar, S., Sastry, N., Redi, M.: Like at first sight: understanding user engagement with the world of microvideos. In: Ciampaglia, G.L., Mashhadi, A., Yasseri, T. (eds.) SocInfo 2017. LNCS, vol. 10539, pp. 237–256. Springer, Cham (2017). https://doi.org/10.1007/978-3-319-67217-5_15

8. Mitchell, V.L.: Knowledge integration and information technology project performance. Mis Q. **30**, 919–939 (2006)

9. Richier, C., Altman, E., Elazouzi, R., Altman, T., Linares, G., Portilla, Y.: Modelling view-count dynamics in YouTube. arXiv preprint arXiv:1404.2570 (2014)

10. Shin, D., He, S., Lee, G.M., Whinston, A.B., Cetintas, S., Lee, K.C.: Enhancing social media analysis with visual analytics: a deep learning approach (2017)

11. Tucker, C.E.: The reach and persuasiveness of viral video ads. Mark. Sci. **34**(2), 281–296 (2014)

12. Yu, H., Xie, L., Sanner, S.: The lifecycle of a YouTube video: phases, content and popularity, pp. 533–542. ICWSM (2015)

Author Index

Printed in the United States
By Bookmasters